BIOLOGY MODULES

Team Co-ordinator
James Torrance

Writing Team
James Torrance
James Fullarton
Clare Marsh
James Simms
Caroline Stevenson

Diagrams by James Torrance

Hodder & Stoughton

A MEMBER OF THE HODDER HEADLINE GROUP

British Library Cataloguing in Publication Data
Torrance, James
  Biology Modules
  I. Title
  574.07

  ISBN 0-340-60097-7

First published 1994
Impression number 10 9   8   7   6   5   4   3   2
Year   1998   1997   1996   1995   1994

Typeset by Litho Link Limited, Welshpool, Powys, Wales
Printed in Great Britain for Hodder & Stoughton
Educational, a division of Hodder Headline Plc, 338 Euston
Road, London NW1 3BH by Bath Press, Avon

Contents

SECTION 1   Cell Biochemistry (module)

1 Molecular structure of cell components   *1*
2 Enzyme action   *16*
3 Experiments relating to cell processes   *32*
4 Chemical pathways   *37*

SECTION 2   Nucleic Acids and Protein Synthesis (half module)

5 Molecular structure of nucleic acids   *49*
6 Genetic code and protein synthesis   *58*
7 Gene expression and gene mutation   *65*

SECTION 3   Genetics (half module)

8 Genetic importance of meiosis   *75*
9 Genetic consequences of chromosomal change   *88*
10 Concepts of genetics   *105*

SECTION 4   Plant Physiological Processes (module)

11 Vital physiological processes   *122*
12 Responses to growth substances   *142*
13 Responses to environmental stimuli   *161*
14 Adaptations to environments   *184*

APPENDIX 1   The genetic code   *201*
INDEX   *202*

PREFACE

Biology Modules will be an invaluable resource for students and teachers. It provides a wealth of material designed to cover a cluster of SCOTVEC modules in biology, which can be used to comprise a complete one-year course, though the reader should not regard the entire content of this book as mandatory to achieve competence in these modules. The book is aimed in particular at school and FE college students who have completed SCE Standard Grade Biology at grade 3 or 4, and can be used as:

- an alternative course to SCE Higher Grade Biology;
- a pre-Higher course for students who would benefit from a preparatory year before tackling SCE Higher Grade Biology or other areas of Higher Education;

- part of a general science course for senior students who may choose one or more modules from the book to add to modules from other science disciplines.

Each chapter consists of a concise set of notes interspersed with Key Questions, to continuously test students' knowledge, and consolidate the learning process.

Each chapter is followed by a varied selection of exercises designed to give practise in, and preparation for, end-of-chapter assessments. These assessments, plus Record of Assessment summary sheets and full answers, are available in the accompanying publication *Biology Modules – Assessment Pack* (ISBN 0340 60098 5) by the same authors.

The authors and publishers would like to thank the **Scottish Vocational Educational Council** for their valuable comments and help in the development of this book.

SECTION 1 : Cell Biochemistry

1 MOLECULAR STRUCTURE OF CELL COMPONENTS

PROTEINS

Proteins are very large molecules. Each is composed of many sub-units called **amino acids** (of which there are twenty different types).

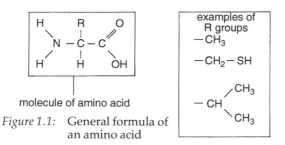

Figure 1.1: General formula of an amino acid

Figure 1.1 shows the general formula of an amino acid molecule. All twenty types possess this combination of **carbon**, **hydrogen**, **oxygen** and **nitrogen** atoms. They differ from one another only by the chemical structure of their **R group** (see figure 1.1); some amino acids for example contain sulphur in their R group. It is the R group that determines an amino acid's individual chemical properties.

PEPTIDE BOND

Two amino acid molecules become joined together by a **peptide bond** which forms as a result of a condensation reaction. This type of reaction

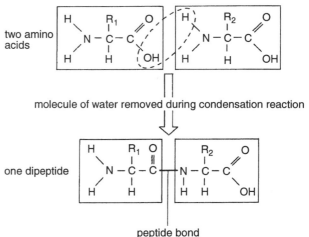

Figure 1.2: Formation of a peptide bond

involves the removal of a molecule of water as shown in figure 1.2. A peptide bond is a type of strong chemical bond. Further peptide bonding leads to the formation of a **polypeptide chain** (see figure 1.3) which often consists of hundreds of amino acids linked together.

HYDROGEN BOND

When hydrogen bonds (which are weak chemical bonds) form between certain amino acids in a polypeptide chain, this results in the chain becoming coiled into a spiral (called a helix) as shown in figure 1.3.

1

Figure 1.3:   Structure of proteins

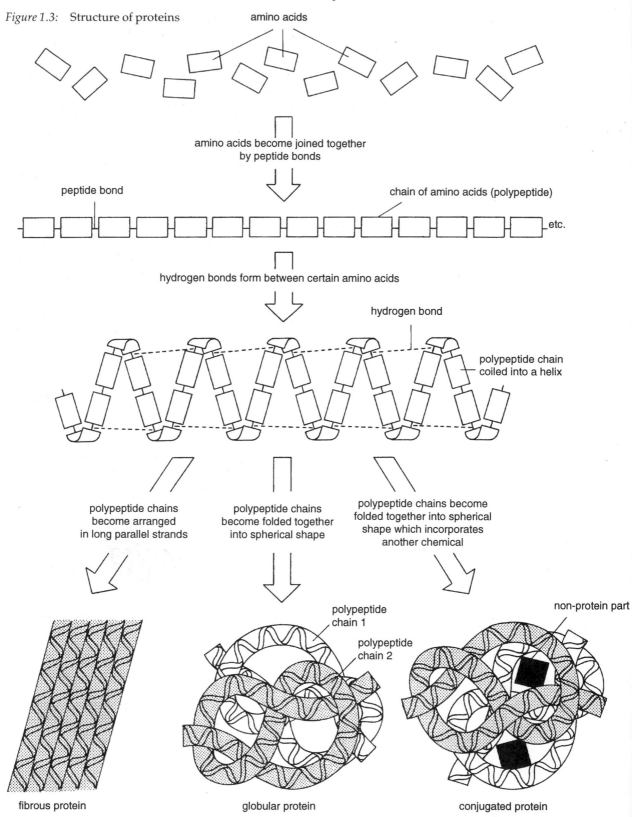

amino acids

amino acids become joined together
by peptide bonds

peptide bond

chain of amino acids (polypeptide)

etc.

hydrogen bonds form between certain amino acids

hydrogen bond

polypeptide chain
coiled into a helix

polypeptide chains
become arranged
in long parallel strands

polypeptide chains
become folded together
into spherical shape

polypeptide chains become
folded together into spherical
shape which incorporates
another chemical

polypeptide
chain 1

polypeptide
chain 2

non-protein part

fibrous protein

globular protein

conjugated protein

FIBROUS PROTEIN

A **fibrous** protein is formed by several spiral-shaped polypeptide molecules becoming linked together in parallel by cross-bridges forming between them. This gives the molecule of structural protein a rope-like structure (see figure 1.3).

Many types of fibrous protein exist, each possessing structural properties which suit it to the role that it plays in the organism's body (see table 1.1).

fibrous protein	structural properties	tissue rich in this protein	function of this protein
elastin	strong and elastic	ligament	flexible attachment of bone to bone
collagen	strong and inelastic	bone	support
keratin	strong and inelastic	hair	protection
actin and myosin	contractile	muscle	movement

Table 1.1   Fibrous proteins

GLOBULAR PROTEIN

A molecule of **globular** protein consists of several polypeptide chains folded together into a roughly spherical shape like a tangled ball of string (see figure 1.3). The exact form that the folding takes depends on the properties of the amino acid R groups on the polypeptide chains.

When sulphur is present in some of the R groups, the polypeptide chains are held together in a certain way by strong bonds (called disulphide bonds) forming between these R groups. The protein molecule's shape may also be maintained by strong bonds involving positive and negative charges and/or weaker hydrogen bonds forming between some pairs of R groups.

Globular proteins are vital components of all living cells. They play several roles, two of which are described below.

Enzymes

All **enzymes** (biological catalysts) are made of globular protein. Each is folded in a particular way to expose an active site which readily combines with a specific substrate (see figure 1.4). Since enzymes speed up the rate of biochemical processes such as photosynthesis, respiration and protein synthesis, they are essential for the maintenance of life.

Structural protein

Globular protein is one of the components which make up the **cell membrane** surrounding a living cell. The other component is phospholipid (see page 10). Although the precise arrangement of these molecules is still unknown, most evidence supports the **fluid-mosaic model** of cell membrane structure as shown in figure 1.5.

Figure 1.4:   Enzyme structure

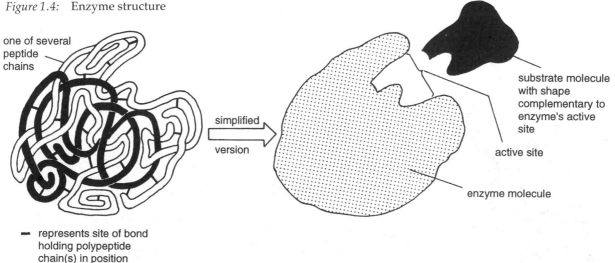

one of several peptide chains

simplified version

substrate molecule with shape complementary to enzyme's active site

active site

enzyme molecule

— represents site of bond holding polypeptide chain(s) in position

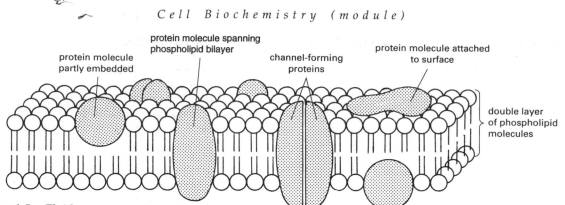

Figure 1.5: Fluid mosaic model of cell membrane

This proposes that the plasma membrane consists of a **fluid layer** of constantly moving phospholipid molecules containing a patchy **mosaic** of protein molecules. These protein molecules vary in size, structure and function. Some, for example, play a purely structural role; others possess a molecular structure able to form pores through which small water soluble molecules can enter or leave the cell.

CONJUGATED PROTEIN

When globular protein is associated with a non-protein chemical, the resultant molecule is called a **conjugated** protein (see figure 1.3).

Haemoglobin (the oxygen-carrying pigment in blood) is an example of a conjugated protein. Figure 1.6 shows it to consist of four polypeptide chains and four non-protein **haem** groups. Each haem has an atom of iron at its centre which can combine reversibly with one molecule of oxygen.

Bonding between the polypeptide chains holds the parts of each molecule together as a stable structure. Further bonding holds the four haem

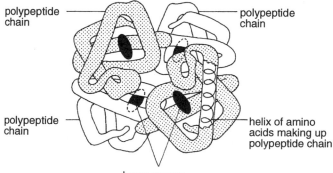

Figure 1.6: Haemoglobin molecule

groups (the reactive sites) in positions which are spaced out from each other. This molecular structure makes haemoglobin ideally suited to pick up and transport oxygen efficiently.

KEY QUESTIONS

1 a) What general name is given to the sub-units of which proteins and polypeptides are composed?
b) How many types of this sub-unit are known to occur in proteins?
c) Identify the FOUR chemical elements common to all amino acids.
d) Which part of an amino acid molecule determines its individual chemical properties?

2 a) What name is given to the type of bond that forms when two amino acid molecules become linked together?
b) Suggest why the reaction which brings about this bonding is called a condensation reaction.
c) What name is given to the chain formed when several amino acids become linked together?

3 a) What type of bond between certain amino acids in a polypeptide chain results in the chain becoming coiled into a helix?
b) Are such chemical bonds stronger or weaker than the type referred to in question 2a?

4 a) Describe TWO ways in which polypeptide chains can become arranged to form a protein.
b) Name TWO types of (i) fibrous (ii) globular protein and for each briefly describe its role.

5 a) What is meant by the term *conjugated protein*?
b) Give ONE example of a conjugated protein and state its function.

CARBOHYDRATES

Carbohydrates (e.g. sugar, starch, glycogen and cellulose) are compounds whose molecules all contain the chemical elements **carbon**, **hydrogen** and **oxygen.**

MONOSACCHARIDES

Simple sugars such as **glucose** are called **monosaccharides** ('single' sugars). The formula for glucose is $C_6H_{12}O_6$. When solid glucose is dissolved in water, its molecules form two types of ring structure as shown in figure 1.7.

Close examination of α-glucose and β-glucose reveals that they differ only by the position of one H and one OH group. They can therefore be represented in a simpler way as shown in figure 1.8. The upper diagrams use numbers to represent the six carbon atoms, while the lower diagrams take an even simpler approach.

Figure 1.7: Two ring forms of glucose

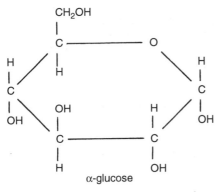

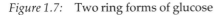

Figure 1.8: Simpler ways of representing glucose molecules

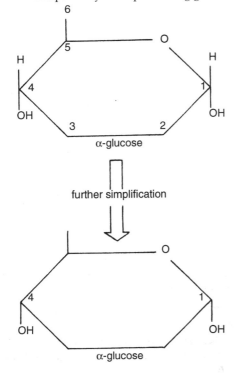

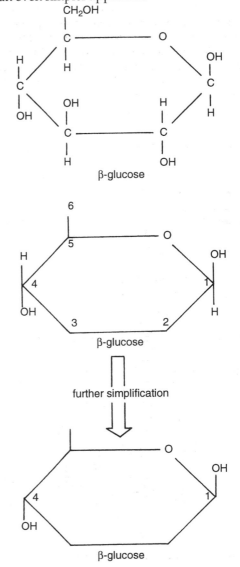

DISACCHARIDES

A **disaccharide** ('double' sugar) is a sugar consisting of two monosaccharide molecules joined together to form a single molecule. Figure 1.9 shows how two molecules of α-glucose become joined together to form **maltose** (a disaccharide).

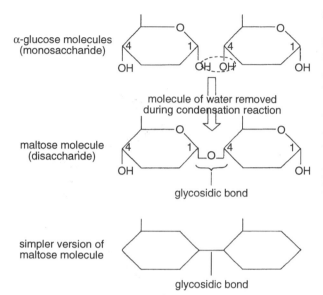

Figure 1.9: Glycosidic bonding in a dissaccharide

This reaction which involves the removal of a molecule of water is called a condensation reaction. The bond formed is known as a **glycosidic bond**. (Its full name is a 1 α-4 glycosidic bond since it involves α-glucose molecules and connects carbon atom 1 of one glucose to carbon 4 of the other.)

POLYSACCHARIDES

A **polysaccharide** is a carbohydrate consisting of many monosaccharide units joined together by glycosidic bonds.

Starch

This carbohydrate consists of a long chain of α-glucose molecules. These are joined together by 1 α-4 glycosidic bonds formed by condensation as in the formation of maltose. Figure 1.10 shows a small part of a starch molecule. The constituent glucose molecules are arranged like a row of coins all in the 'heads up' position.

The complete molecule may contain a thousand or more glucose units. The molecules of some forms of starch bear side branches. These occur about once every 25 glucose units, see figure 1.11.

The glucose units in a starch molecule coil into a **helix** (see figure 1.12) which is held together by hydrogen bonds. Several starch molecules become gathered together and stored as a starch **granule** (grain). Starch molecules are large and insoluble in water. This means that they cannot diffuse out of the cell and that they do not affect the cell's osmotic balance or metabolism. The molecular structure of starch is therefore ideally suited to its role of **food storage.**

Glycogen

This polysaccharide is the storage carbohydrate of animals. Figure 1.11 shows that it has a molecular structure similar to the branched form of starch except that its side arms emerge about every ten glucose units. Like starch it is insoluble, which makes it an ideal storage material. It exists as tiny **granules** particularly abundant in liver cells.

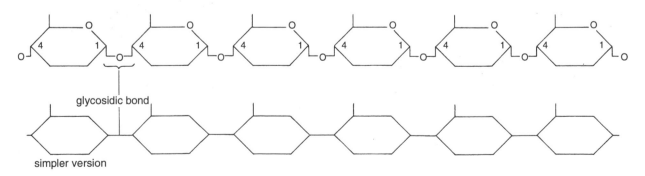

Figure 1.10: Small part of an unbranched starch molecule

small part of a
branched starch
molecule

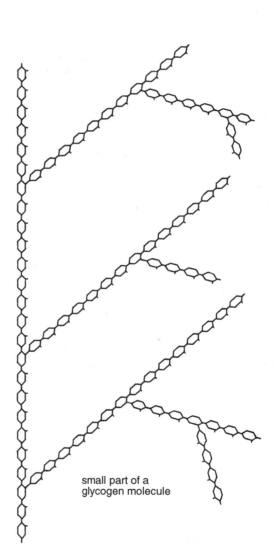

small part of a
glycogen molecule

Figure 1.11:   Two branched polysaccharides

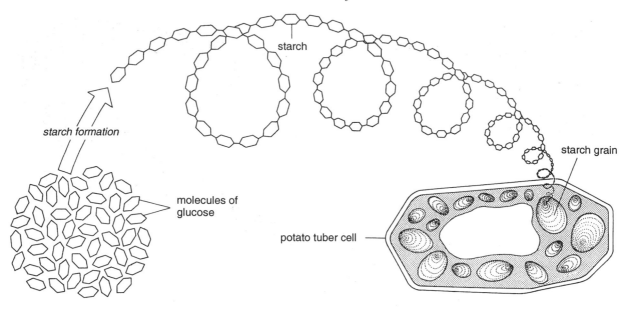

Figure 1.12:   Starch, the storage carbohydrate in
                 plants

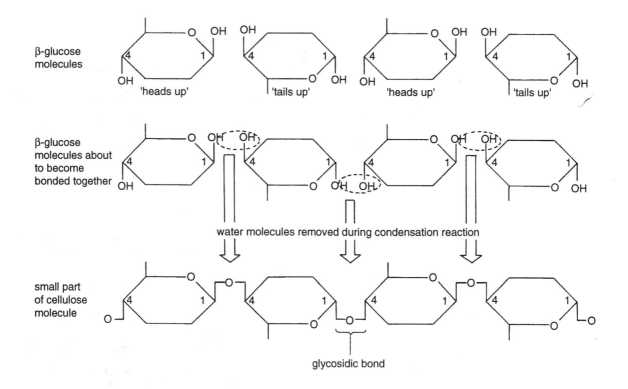

Figure 1.13:   Formation of part of a cellulose molecule

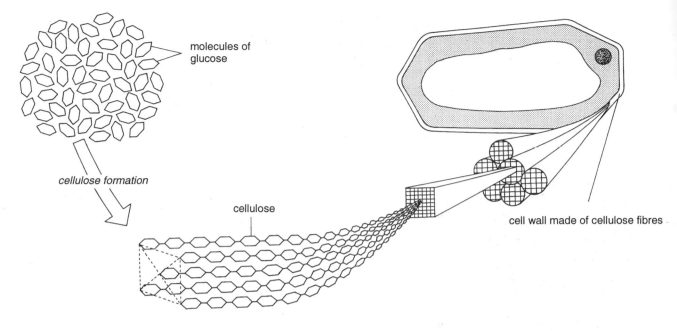

Figure 1.14: Cellulose, the structural carbohydrate in plants

Cellulose

Cellulose is a polysaccharide consisting of β-glucose molecules (see figure 1.7) linked together by 1 β-4 glycosidic bonds. This form of bonding results in the constituent glucose molecules being arranged along the cellulose molecule like a row of coins in alternate 'heads up' and 'tails up' positions (see figure 1.13).

This arrangement of glucose molecules does not coil up or bear branches and the cellulose molecule remains like a **flat ribbon.** Hydrogen bonds form between adjacent ribbons and hold them together as tiny **fibres.** These become grouped into larger fibres which are laid down in basket-like layers to form the wall of every plant cell (see figure 1.14). This molecular arrangement gives a strong, rigid yet slightly elastic structure which is ideal as the main component of the cell wall.

KEY QUESTIONS

1 Name the THREE chemical elements of which all carbohydrates are composed.

2 **a)** Give the chemical formula of glucose.
   b) (i) Name the TWO molecular forms of glucose that occur when glucose powder is dissolved in water. (ii) Identify the difference between these two forms of glucose.

3 Draw a simple diagram of a molecule of a named monosaccharide, disaccharide and polysaccharide.

4 **a)** Describe the molecular structure of both forms of starch in terms of monosaccharide units.
   b) Name the type of bond that holds the units together in a starch molecule.
   c) What is the function of starch in a plant cell?

5 **a)** Describe the molecular structure of cellulose in terms of monosaccharide units.
   b) What is the function of cellulose in a plant cell?

Lipids

These include fats, oils and phospholipids. Like carbohydrates, lipids contain the chemical elements **carbon**, **hydrogen** and **oxygen**. However the proportion of oxygen in a lipid molecule is much smaller than that in a molecule of carbohydrate.

Simple lipids

Fats (solid at room temperature) and **oils** (liquid at room temperature) are simple lipids. Each molecule of a simple lipid consists of one **glycerol** molecule combined to three **fatty acid** molecules.

The molecular structure of glycerol (see figure 1.15) is constant from lipid to lipid. Fatty acids however vary in molecular structure. A fatty acid can be represented by the general formula shown in figure 1.15 (where n varies from one fatty acid to another) and takes the form of a long hydrocarbon chain.

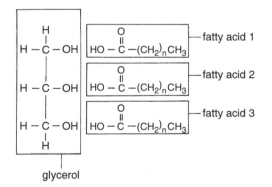

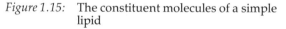

Figure 1.15: The constituent molecules of a simple lipid

Formation of a simple lipid

Each of a glycerol molecule's three —OH groups combines with a molecule of fatty acid to form one molecule of simple lipid. This involves a condensation reaction during which three molecules of water are removed as shown in figure 1.16.

Each fatty acid molecule is linked to glycerol by a **covalent bond** which is a type of strong chemical bond. All the other C—H, C—C and C—O bonds in a lipid are also covalent bonds.

Figure 1.16: Formation of a simple lipid molecule

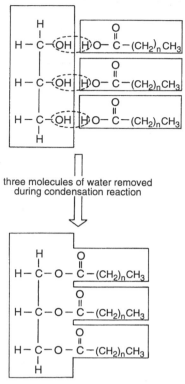

three molecules of water removed during condensation reaction

molecule of lipid

Figure 1.17 shows a simpler way of representing one molecule of lipid.

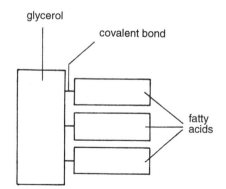

Figure 1.17: Simplified diagram of a lipid

Phospholipid

A typical **phospholipid** molecule contains two molecules of fatty acid linked to one of glycerol as before. However the third position on the glycerol molecule is occupied by a **phosphate** group attached to the opposite side from the fatty acid molecules (see figure 1.18).

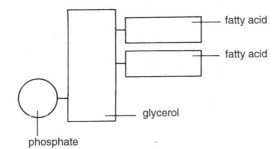

Figure 1.18: Molecule of phospholipid

The two ends of a molecule of phospholipid possess different properties. The phosphate ('head') end of the molecule is **hydrophilic** ('water-loving') and therefore soluble in water. The fatty acid ('tail') end of the molecule is **hydrophobic** ('water-hating') and therefore insoluble in water. This is summarised in figure 1.19 which shows a simpler way of representing a phospholipid molecule.

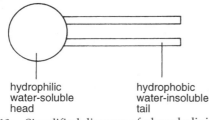

Figure 1.19: Simplified diagram of phospholipid

Cell membrane

Phospholipids are an especially important group of lipids. Along with molecules of protein, they are the basic building blocks of all **cell membranes.**

The fact that one end of a phospholipid molecule is hydrophilic and the other end is hydrophobic ideally suits the molecule to its role as a basic constituent of the cell membrane. If it is in the company of a large number of similar phospholipid molecules, the group becomes arranged into a **bilayer** (see figure 1.20).

The water-soluble hydrophilic heads lie to the outside where they form hydrogen bonds with the surrounding water molecules; the water-insoluble hydrophobic tails point inwards attracted to those of the other layer.

This arrangement of phospholipid molecules is fluid yet at the same time forms a fairly stable and effective **barrier** round the cell and the cell organelles. It prevents the passage of water and water-soluble molecules through it. (Transport of such molecules through the membrane involves the protein molecules — see page 4).

KEY QUESTIONS

1 a) Name the THREE chemical elements present in all lipids.
b) Which of these makes up a smaller proportion of the molecular weight of a lipid compared with that of a carbohydrate?
c) Name TWO types of simple lipid.

2 a) To what other type of sub-unit are the fatty acids in a simple lipid molecule combined?
b) How many fatty acid molecules are linked to one molecule of this other sub-unit?
c) Name the type of chemical bond that holds the constituents of a simple lipid molecule together.

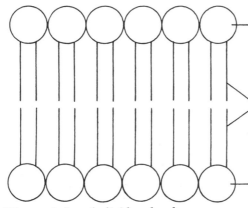

Figure 1.20: Bilayer of phospholipid molecules

3 a) Give ONE difference between a molecule of simple lipid and one of phospholipid.
b) Which end of a phospholipid molecule is hydrophobic? What does this mean?
c) Which end of a phospholipid molecule is hydrophilic? What does this mean?
d) (i) Describe the form that a mass of phospholipid molecules takes in a cell membrane. (ii) Of what value is such a structure to the cell?

EXERCISES

1 Match the terms in list **X** with their descriptions in list **Y**.

list X	list Y
1) amino acid	**a)** strong chemical link joining adjacent glucose molecules in large carbohydrate molecules
2) carbo-hydrate	**b)** large molecule composed of amino acids linked together into one or more polypeptide chains
3) fatty acid	**c)** one of twenty types of organic compound which are the basic building blocks of proteins
4) glucose	**d)** general name for a large group of compounds which includes fats and oils
5) glycerol	**e)** strong chemical link joining adjacent amino acids in a polypeptide chain
6) glycosidic bond	**f)** general name for a compound composed of one or more monosaccharide units
7) hydrogen bond	**g)** compound which bears a long hydrocarbon chain and is a basic component of a lipid
8) lipids	**h)** 6-carbon sugar which is the basic building block of many polysaccharides
9) peptide bond	**i)** molecule with which three fatty acids combine to form a simple lipid
10) protein	**j)** weak chemical bond holding a polypeptide chain in a coil by linking adjacent amino acids in the helix.

2 Which of the following chemical elements is **always** present in protein but absent from carbohydrate?
A carbon
B hydrogen
C oxygen
D nitrogen
(Choose ONE correct answer only.)

3 Match the terms in list **X** with their descriptions in list **Y**.

list X	list Y
1) cellulose	**a)** molecule consisting of polypeptide chains arranged like a ball of string
2) conjugated protein	**b)** water-loving head end of phospholipid molecule
3) fibrous protein	**c)** type of lipid molecule possessing phosphate group in place of a fatty acid
4) globular protein	**d)** water-hating tail end of phospholipid molecule
5) glycogen	**e)** large insoluble carbohydrate molecule which is basic component of plant cell wall
6) hydrophilic part	**f)** chain-like molecule composed of several amino acids
7) hydro-phobic part	**g)** large insoluble molecule which acts as the storage carbohydrate in plant cells
8) phospho-lipid	**h)** molecule consisting of long parallel polypeptide chains arranged like a rope
9) polypeptide	**i)** large insoluble carbohydrate molecule stored as granules in animal cells
10) starch	**j)** molecule consisting of polypeptide chains folded into a ball and associated with a non protein chemical

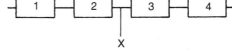

Figure 1.21

4 Figure 1.21 shows a small portion of a polypeptide chain.
a) What general name is given to the numbered sub-units in the diagram?

b) Draw a simple diagram of the general chemical formula which represents all such sub-units.

c) Identify the chemical bond found at X in the diagram.

d) What general name is given to the type of compound which consists of one or more polypeptide chains?

5 Figure 1.22 shows three different types of protein.

a) Which of these is:

(i) a globular protein?

(ii) a conjugated protein?

(iii) a fibrous protein?

(iv) composed of only one polypeptide chain?

b) (i) Which diagram represents collagen?

(ii) Explain how you arrived at your answer.

(iii) Describe how collagen's structure is ideally suited to its function.

c) (i) Which diagram represents a molecule of myoglobin, a red oxygen-carrying pigment found in vertebrate muscle?

(ii) Explain your choice of answer.

(iii) Describe how the molecular structure of myoglobin enables it to carry out its function efficiently.

d) (i) Which diagram represents a molecule of enzyme?

(ii) Explain how you arrived at your answer.

(iii) Make a simple sketch to show the possible appearance of a molecule of the substrate associated with this enzyme.

6 Figure 1.23 shows the possible arrangement of the molecules in a cell membrane.

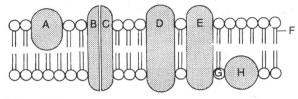

Figure 1.23

a) Identify by their letters those molecules in the diagram which are composed of one or more polypeptide chains.

b) (i) Give the general name of the sub-units of which the polypeptide chains are, in turn, composed.

(ii) Draw a diagram of FIVE of these sub-units to show how they would be arranged in a polypeptide molecule.

(iii) Name the type of bond that connects one of these sub-units to the next in the polypeptide chain.

c) (i) Which lettered structures in the diagram enclose a narrow channel?

(ii) What is the function of such a channel?

d) Give ONE way in which the structure of molecule D in the diagram differs from that of a molecule of elastin.

e) i) Identify molecule F.

(ii) Make a more detailed diagram of it to show the THREE types of molecule of which it is composed. Name them in your diagram.

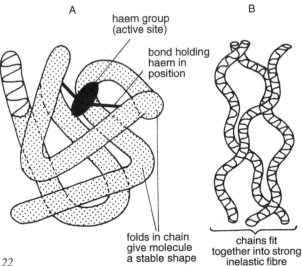

Figure 1.22

A — haem group (active site); bond holding haem in position; folds in chain give molecule a stable shape

B — chains fit together into strong inelastic fibre

C — active site

(iii) Name the type of chemical bond that holds the constituents of molecule F together.

f) (i) Describe the properties of molecule F that make it form a bilayer when in the company of other similar molecules and water.

(ii) Explain why this molecular arrangement is ideal as an essential part of a cell membrane.

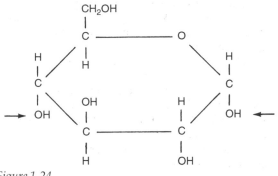

Figure 1.24

7 Figure 1.24 represents a molecule of carbohydrate.

a) Which chemical elements are represented by the symbols C, H and O?

b) Give the chemical formula of this carbohydrate.

c) (i) Draw a simpler version of the diagram which includes the two OH groups indicated by arrows.

(ii) Using this simpler style, draw a diagram to show two of these monosaccharide units joined together.

(iii) What general name is given to this type of carbohydrate composed of two monosaccharide units?

(iv) What name is given to the bond linking these two sub-units together?

8 Figure 1.25 is a very simple representation of a small portion of a starch molecule.

Figure 1.25

a) Make a similar diagram of a small part of a molecule of the other form of starch by drawing about 70 sub-units and their bonds.

b) State ONE way in which the structure of a molecule of glycogen differs from the starch molecule that you have drawn.

c) Name the monosaccharide sub-unit which makes up (i) starch (ii) glycogen.

d) Name the chemical bond that joins the sub-units in starch and in glycogen.

e) In what way would the structure of a molecule of starch in a starch grain differ from the way that it is represented in the diagram?

f) What function does starch perform in a plant cell?

g) Which feature of a starch molecule prevents it from

(i) leaking out of the cell?

(ii) affecting the cell's osmotic balance?

9 Figure 1.26 is an incomplete representation of a molecule of simple lipid.

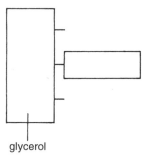

glycerol

Figure 1.26

a) Draw a completed version of the diagram and label the constituent parts.

b) Name the type of bond which links one constituent molecule to another in a lipid.

c) (i) Give ONE way in which a phospholipid differs from a simple lipid in molecular structure.

(ii) Identify the structural role played by phospholipids in all living cells.

10 Figure 1.27 shows a small part of a starch molecule.

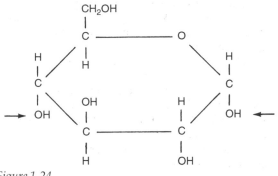

Figure 1.27

a) Using the same style, draw a diagram to show how five sub-units would be arranged in a small part of a cellulose molecule.

b) Name the type of monosaccharide sub-unit of which a cellulose molecule is composed.

c) Name the bond which links the sub-units together in a cellulose molecule.

d) Sometimes the sub-units in a starch molecule are likened to a row of coins all arranged in the 'heads up' position. On this basis how would you describe the arrangement of the sub-units in a cellulose molecule?

e) Rewrite the following sentences selecting only the correct answer at each choice.

Cellulose is a $\left\{\begin{array}{l}\text{monosaccharide}\\\text{polysaccharide}\end{array}\right\}$ molecule

which $\left\{\begin{array}{l}\text{remains unbranched.}\\\text{becomes branched.}\end{array}\right\}$ In addition, it

$\left\{\begin{array}{l}\text{becomes coiled}\\\text{remains straight}\end{array}\right\}$ and, in association with four

other similar cellulose molecules, forms a tiny

$\left\{\begin{array}{l}\text{fibre}\\\text{granule}\end{array}\right\}$ held together by $\left\{\begin{array}{l}\text{covalent}\\\text{hydrogen}\end{array}\right\}$ bonds.

f) Describe the properties possessed by a plant cell wall which are largely due to its component cellulose fibres.

11 Decide whether each of the following statements is **true** or **false** and then use T or F to indicate your choice. Where a statement is false, give the word that should have been used in place of the word in **bold print.**

a) The type of link joining adjacent amino acid molecules in a polypeptide chain is called a **glycosidic** bond.

b) The type of carbohydrate stored in granules in animal cells is **glycogen.**

c) Starch and cellulose are examples of **monosaccharide** carbohydrate molecules.

d) A molecule of simple lipid consists of fatty acids and **glucose.**

e) A molecule of enzyme is normally composed of **globular** protein.

2 ENZYME ACTION

ACTIVATION ENERGY

The **rate** of a chemical reaction is indicated by the amount of chemical change that occurs per unit time. Such change may involve the joining together of simple molecules into more complex ones or the splitting of complex molecules into simpler ones. In either case the energy needed to make the chemical reaction go is called its **activation energy.**

This often takes the form of heat energy and the reaction only proceeds at a high rate if the chemicals involved are raised to a high temperature (see figure 2.1). However living cells cannot tolerate such high temperatures. So how is this problem overcome in a living system?

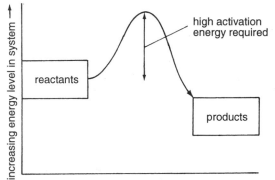

Figure 2.1: Uncatalysed reaction

BIOLOGICAL CATALYST

A **catalyst** is a substance which increases the rate of a chemical reaction by lowering the amount of activation energy needed to make the reaction proceed (see figure 2.2).

Enzymes are biological catalysts. They speed up the rate of biochemical reactions by lowering their activation energy requirements. Enzymes therefore make it possible for reactions to occur in cells at temperatures which would otherwise be

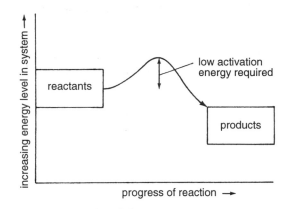

Figure 2.2: Catalysed reaction

too low for chemical molecules to react at a fast rate. In the absence of enzymes, biochemical reactions would proceed so slowly that life as we know it would cease to exist.

Since the molecules of an enzyme do not undergo any permanent change in structure as a result of the reaction that they catalyse, they can perform their role **repeatedly.** A relatively small amount of enzyme is therefore able to catalyse a large amount of substrate into product.

Many different enzymes are made by and are present in all living cells. Every biochemical reaction that occurs in a living organism is catalysed by an enzyme. Some of these reactions occur inside cells and are controlled by **intracellular** enzymes (e.g. cytochrome oxidase plays a key role in aerobic respiration — see page 39); other reactions occur outside cells under the influence of **extracellular** enzymes (e.g. pepsin promotes the digestion of protein in the stomach).

EFFECT OF AMYLASE ON STARCH

Look at the experiment shown in figure 2.3. The conditions in the two test tubes are identical at the start except that tube A receives the enzyme **amylase** whereas B (the control) receives water.

From the results it is concluded that in tube A the enzyme amylase has promoted the breakdown

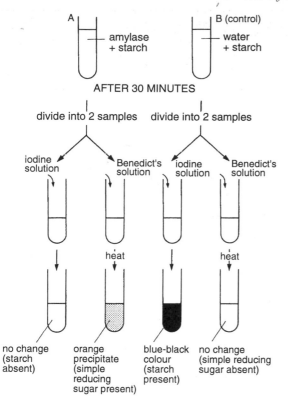

Figure 2.3: Action of amylase

of starch to simple sugar (maltose). In tube B, the control, which lacks the enzyme, no detectable reaction has occurred.

The substance upon which an enzyme acts is called the **substrate.** The substance produced as a result of the reaction is called the **end product.** The reaction being promoted in this experiment can therefore be summarised by the following word equation:

$$\text{starch} \xrightarrow{\text{amylase}} \text{simple sugar (maltose)}$$
$$\text{(substrate)} \quad \text{(enzyme)} \quad \text{(end product)}$$

STRUCTURE

All enzyme molecules are made of **protein.** Most enzymes are simple globular proteins and consist of one or more polypeptide chains folded together like a tangled ball of string (see page 3). A few enzymes are made of conjugated protein (i.e. globular in association with a non-protein part).

At some point on the surface of an enzyme molecule there is an **active site** which has a particular shape. This is determined by the chemical structure of and type of bonding between the polypeptide chains which make up the enzyme molecule.

MECHANISM OF ACTION

An enzyme is able to act on only one type of substance (its substrate) since this is the only substance whose molecules exactly fit the

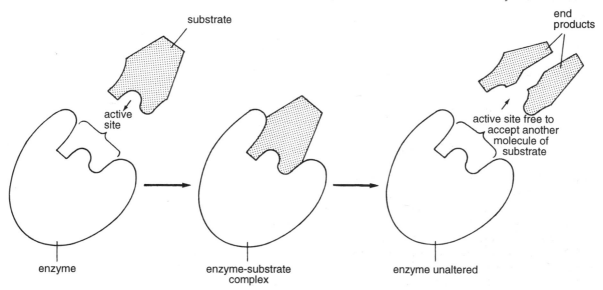

Figure 2.4: Lock and key mechanism of enzyme action (degradation of complex substrate)

Figure 2.5: Lock and key mechanism of enzyme action (synthesis of complex product)

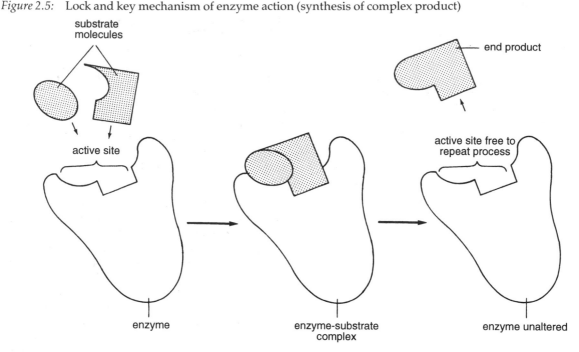

enzyme's active site. The enzyme is said therefore to be **specific** to its substrate and the substrate's molecular shape is said to be **complementary** to the enzyme's active site.

The **lock-and-key** mechanism is the name given to the means by which enzymes are thought to operate. This hypothesis proposes that the substrate combines with the enzyme at its active site in a precise way just as a lock and key fit together. Figure 2.4 shows how the two combine briefly as an enzyme-substrate complex, allowing the reaction to occur. The end products escape from the active site leaving the enzyme **unaltered** and free to combine with another molecule of substrate.

Some enzymes promote the **breaking-down** of complex molecules to simpler ones (see figure 2.4); others promote the **building-up** of complex molecules from simpler ones (see figure 2.5).

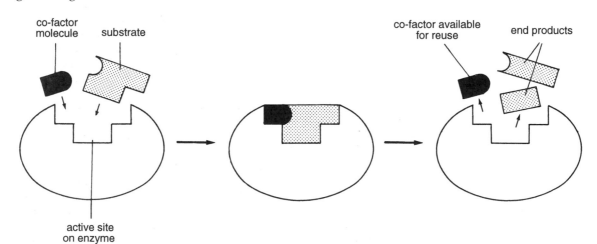

Figure 2.6: Action of enzyme co-factor

CO-FACTORS

Many enzymes require the presence of a non-protein substance called a **co-factor** to function efficiently and bring about their catalytic effect (see figure 2.6).

In some cases the co-factor is thought to play its part by enabling the substrate molecule to fit correctly at the enzyme's active site. Other co-factors provide the energy needed to drive the chemical reaction.

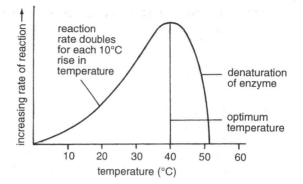
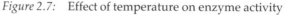

Figure 2.7: Effect of temperature on enzyme activity

At very low temperatures, enzyme molecules are **inactive** but undamaged. At low temperatures, enzyme and substrate molecules move around **slowly** in their surrounding medium. They meet only rarely and the rate of enzyme activity remains low.

As the temperature increases, the two types of molecule move about at a **faster rate** and more molecular collisions occur. A greater number of enzyme-substrate complexes are formed and the rate of reaction increases. However this trend does not continue beyond temperatures of around 40°C (see figure 2.7).

DENATURATION

Increase in temperature makes an enzyme's atoms **vibrate.** At temperatures above 40°C, its atoms vibrate so much that some of the chemical bonds (e.g. hydrogen bonds) that hold the enzyme together in its specific shape, start to break. The enzyme's constituent polypeptide molecules start to unravel and soon its active site is lost. An enzyme in this state is said to be **denatured.**

Since 40°C (approximately) is the temperature at which most molecular collisions occur yet denaturation has barely begun, this is the temperature at which the reaction works best. It is called the enzyme's **optimum** temperature.

Beyond 40°C, as molecules of enzyme become denatured, the rate of the reaction decreases rapidly. At temperatures of about 55–60°C, enzyme activity is found to have come to a complete halt. This is because all of the enzyme's molecules have become denatured. (Exceptions to this rule are algae living in hot springs and thermophilic bacteria, whose unusual enzymes are denatured only by very high temperatures.)

<table>
<tr><td>

KEY QUESTIONS

</td></tr>
</table>

1 a) What is meant by the term *rate of reaction?*
  b) Why is an enzyme described as a biological catalyst?
  c) What is the difference between an extracellular and an intracellular enzyme?

2 a) Of which type of organic substance are enzymes composed?
  b) What TWO factors determine the shape of an enzyme's active site?

3 a) What name is given to the type of substance upon which an enzyme acts?
  b) Why is the enzyme said to be specific in its relationship with its substrate?
  c) According to the lock-and-key hypothesis, how is an enzyme thought to act?
  d) What name is given to a non-protein substance that must be present for some enzymes to function efficiently?

FACTORS AFFECTING ENZYME ACTION

To function efficiently an enzyme requires a suitable temperature, a suitable pH and an adequate supply of substrate. Inhibitors may bring the enzyme's action to a partial or even complete halt.

EFFECT OF TEMPERATURE
(See chapter 3 for practical investigation)

The graph in figure 2.7 summarises the general effect of temperature on enzyme activity.

1 Salivary amylase is an enzyme which digests starch in the human mouth.
   a) (i) Compare its rate of action at 5°C and 20°C.
   (ii) Explain the difference in rate in terms of behaviour of enzyme and substrate molecules.
   b) (i) Compare the rate of the reaction at 40°C and 60°C.
   (ii) Explain the difference in rate in terms of the molecules.

2 **a)** Which temperature given in question 1 is closest to the optimum for most enzymes?
   b) (i) Suggest a temperature at which an enzyme is inactive but capable of activity if the temperature changes.
   (ii) Explain your answer.

EFFECT OF pH
(See chapter 3 for practical investigation)

The symbol **pH** refers to the concentration of **hydrogen ions** (H^+) present in a solution. The greater the number of hydrogen ions present, the lower the pH. A change of one 'unit' on the pH scale involves a ten-fold increase or decrease in hydrogen ion concentration. A change of two units involves a hundred-fold increase or decrease and so on. Thus what may seem like a small change in pH is really a relatively large change in hydrogen ion concentration in the solution.

The shape of an enzyme molecule is partly due to the presence of hydrogen bonds which hold its constituent polypeptide chain(s) together. Exposure of an enzyme to an extreme of pH (such as the very high concentration of hydrogen ions present in a concentrated acid) normally results in the bonds being broken and the enzyme becoming denatured.

NEUTRALISATION OF ACTIVE SITE

Change in pH can have a further adverse effect on some enzymes. The successful formation of an enzyme-substrate complex often depends on the active site and the substrate's complementary surface having **opposite charges** which attract one another (see figure 2.8).

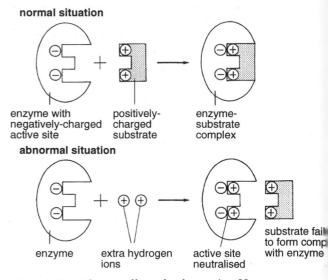

Figure 2.8:   Adverse effect of a change in pH

If negative charges on the enzyme's active site are neutralised by extra H^+ ions (e.g. from acid added to the surrounding medium) then the enzyme will fail to bind to its substrate.

OPTIMUM

Each enzyme works best at a particular pH (its **optimum** pH) as shown in figure 2.9. Many enzymes function within a working pH range of about 5–9 with an optimum at pH 7 (neutral). However there are exceptions. Pepsin secreted by the stomach's gastric glands works best in strongly acidic conditions of pH 2; alkaline phosphatase which plays a role in bone formation works best at pH 10.

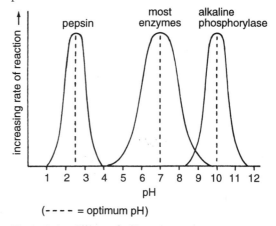

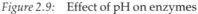

Figure 2.9:   Effect of pH on enzymes

1 a) To what does the symbol pH refer?
b) Name the class of chemicals whose pH is always: (i) below (ii) above number 7 on the pH scale.
c) Which of the two classes of chemical in **b)** has the higher hydrogen ion concentration?
d) Why does exposure to one of these chemicals normally result in denaturation of an enzyme molecule?

2 Describe a further adverse effect that the presence of extra hydrogen ions in the surrounding medium can have on some enzymes.

3 a) What is meant by the term *optimum pH* with respect to enzyme activity?
b) (i) Is this optimum value found to be constant for all enzymes? (ii) Explain your answer.

EFFECT OF ENZYME CONCENTRATION

Each enzyme-substrate complex formed during an enzyme-controlled reaction exists for only a brief

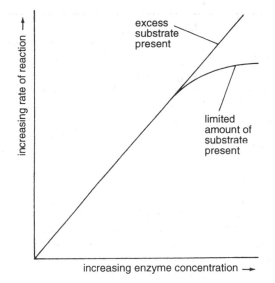

Figure 2.10: Effect of increasing enzyme concentration

moment. Following the release of the products, the enzyme's active site becomes free allowing it to combine with another molecule of substrate, and so on.

The number of substrate molecules which can be acted upon by an enzyme molecule in a given time

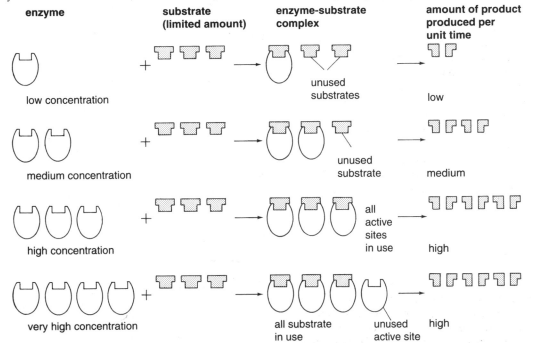

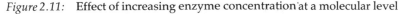

Figure 2.11: Effect of increasing enzyme concentration at a molecular level

is called its **turnover number.** This varies from about 4000 substrate molecules per second in the case of catalase, to around 100 per second for the slowest enzyme.

An increase in concentration (i.e. number of molecules) of **enzyme** results in an increase in rate of reaction since more and more substrate molecules are being acted upon. This relationship is summarised by the graph in figure 2.10.

This upward trend continues as a straight line provided that excess substrate is present. However if the amount of substrate is limited, then the graph levels off since some enzyme molecules fail to find substrate molecules to act upon.

The effect of increasing enzyme concentration is summarised at molecular level in a simplified way in figure 2.11.

EFFECT OF SUBSTRATE CONCENTRATION

The graph in figure 2.12 shows the effect of increasing **substrate** concentration on the rate of an enzyme-controlled reaction for a limited amount of enzyme.

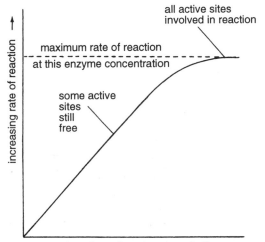

Figure 2.12: Effect of increasing substrate concentration

At low concentrations of substrate the reaction rate is low since there are too few substrate molecules present to make maximum use of all the active sites on the enzyme molecules.

An increase in substrate concentration results in an increase in reaction rate since more and more active sites become involved.

This upward trend in the graph continues as a straight line until a point is reached where further increase in substrate concentration fails to make the reaction go any faster. At this point all the active sites are occupied (the enzyme concentration has become the limiting factor). The graph levels off since there are now more substrate molecules present than there are free active sites with which to combine. The effect of increasing substrate concentration is summarised at molecular level in a simplified way in figure 2.13.

KEY QUESTIONS

1 **a)** What is meant by the term *turnover number* with respect to an enzyme?
   b) Give an example of turnover number for a named enzyme.

2 **a)** In general, what effect does an increase in concentration of enzyme have on reaction rate when the substrate is present in excess? Explain why.
   b) What effect does an increase in concentration of enzyme have on reaction rate when a limited amount of substrate is present? Explain why.

3 **a)** What effect does an increase in concentration of substrate have on reaction rate when the enzyme is present in excess? Explain why.
   b) What effect does an increase in concentration of substrate have on reaction rate when a limited amount of the enzyme is present? Explain why.

EFFECT OF INHIBITORS
(See chapter 3 for practical investigation)

An **inhibitor** is a substance which decreases the rate of an enzyme-controlled reaction and may even bring it to a halt. Inhibitors can be divided into two types: **competitive** and **non-competitive.**

COMPETITIVE INHIBITORS

Molecules of a competitive inhibitor compete with molecules of the substrate for the active sites on

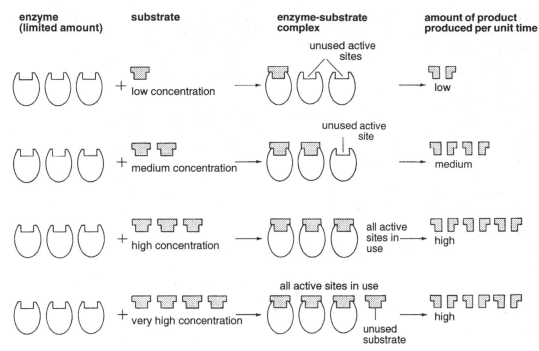

| enzyme (limited amount) | substrate | enzyme-substrate complex | amount of product produced per unit time |

Figure 2.13: Effect of increasing substrate concentration at a molecular level

the enzyme. The inhibitor is able to do this because its molecular structure is similar to that of the substrate and it can attach itself to the enzyme's active site as shown in figure 2.14.

Since active sites **blocked** by competitive inhibitor cannot become occupied by substrate molecules, the rate of the reaction is reduced.

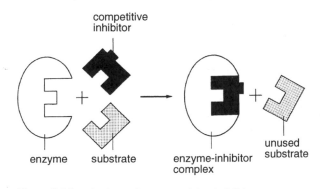

Figure 2.14: Action of a competitive inhibitor

Effect of concentration of competitive inhibitor

For a limited amount of substrate and enzyme, an increase in concentration of competitive inhibitor has the effect on reaction rate shown in figure 2.15.

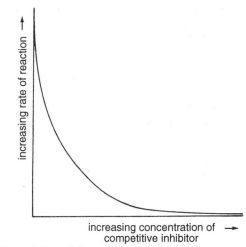

Figure 2.15: Effect of increasing concentration of competitive inhibitor

At low concentrations of inhibitor the reaction rate is high since few active sites are blocked by the inhibitor and substrate molecules are having no difficulty finding free active sites on the enzyme molecules.

However as the concentration of competitive inhibitor increases, the reaction rate decreases owing to the reduced number of unblocked active sites available to substrate molecules.

NON-COMPETITIVE INHIBITORS

These do not combine with the enzyme's active site. Instead a non-competitive inhibitor becomes attached to some other region of the enzyme molecule. This results in the active site being **altered indirectly** as shown in figure 2.16. The substrate is therefore unable to combine with the enzyme.

Cyanide acts in this way by attaching itself to the non-protein iron part of the enzyme cytochrome oxidase. This inhibits the process of aerobic respiration (see page 39). Heavy metals such as mercury and copper are further examples of non-competitive inhibitors.

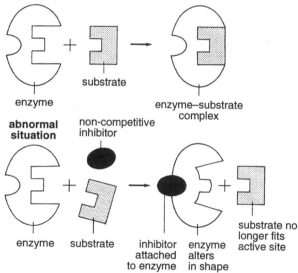

normal situation

substrate

enzyme

enzyme–substrate complex

abnormal situation

non-competitive inhibitor

enzyme   substrate   inhibitor attached to enzyme   enzyme alters in shape   substrate no longer fits active site

Figure 2.16:   Action of a non-competitive inhibitor

Effect of increasing substrate concentration

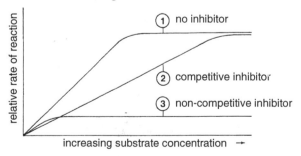

relative rate of reaction

① no inhibitor

② competitive inhibitor

③ non-competitive inhibitor

increasing substrate concentration →

Figure 2.17:   Comparative effects of increasing substrate concentration in the presence of inhibitors

The graph in figure 2.17 compares the effect of increasing substrate concentration on rate of reaction for a limited amount of enzyme affected by limited amount of inhibitor.

In line 1 (the control), increase in substrate concentration brings about an increase in reaction rate until a point is reached where all the active sites on the enzyme molecules are occupied (see also figure 2.12) and then the graph levels off.

In line 2, increase in substrate concentration brings about a gradual increase in reaction rate. Although the **competitive inhibitor** is competing for and occupying some of the enzyme's active sites, the true substrate is also occupying some of the sites. As substrate molecules increase in concentration and outnumber those of the competitive inhibitor, more and more sites become occupied by substrate rather than inhibitor molecules. The reaction rate continues to increase until all the active sites are occupied (almost all of them by substrate).

In line 3, most of the enzyme molecules have been altered (at some position other than the active site) by the **non-competitive inhibitor** and rendered inactive. However a few enzyme molecules remain unaffected and the reaction proceeds at a low rate. Increase in substrate concentration fails to increase reaction rate since the few active sites that are operational are already working at maximum capacity.

SUMMARY

The degree of inhibition of an enzyme by a competitive inhibitor is affected by both the concentration of the inhibitor and the concentration of the substrate present. The degree of inhibition of an enzyme by a non-competitive inhibitor depends on the concentration of the inhibitor only.

IRREVERSIBLE INHIBITORS

Some inhibitors are described as **irreversible** because they become attached to the enzyme's active site permanently and cannot be dislodged. Examples include certain types of insecticide and nerve gas both of which combine with the enzyme **cholinesterase.** Under natural circumstances this enzyme plays a vital role in the transmission of impulses from nerve cell to nerve cell. Its inhibition normally leads to the death of the affected animal.

KEY QUESTIONS

1 Define the term *inhibitor* with respect to enzymes.

2 a) What property of a competitive inhibitor enables it to compete with the substrate?
b) What effect does an increase in concentration of competitive inhibitor have on reaction rate when the amount of substrate present is limited? Explain why.
c) What effect does an increase in concentration of substrate have on reaction rate when the amount of competitive inhibitor is limited? Explain why.

3 Describe how a non-competitive inhibitor acts on an enzyme.

4 Explain why nerve gas works as a very effective poison.

EXERCISES

1 Match the terms in list **X** with their descriptions in list **Y**.

list X	list Y
1) active site	a) substance formed as a result of an enzyme acting on its substrate
2) co-factor	b) process by which an enzyme's structure is changed and its active site is destroyed
3) digestion	c) biological catalyst composed of globular protein
4) denatur-ation	d) building-up of large complex molecules from simpler ones by an enzyme-controlled reaction
5) end product	e) region on an enzyme where the complementary surface of the substrate becomes attached
6) enzyme	f) substance upon which an enzyme acts resulting in the formation of an end product
7 inhibitor	g) non-protein substance needed by some enzymes to bring about their catalytic effect
8 lock-and-key mechanism	h) substance which decreases or halts the rate of an enzyme-controlled reaction
9) substrate	i) breakdown of large complex molecules to simpler ones by an enzyme-controlled reaction
10) synthesis	j) hypothesis that proposes that substrate and enzyme molecules fit together in a precise and specific way

2 Hydrogen peroxide is a chemical which slowly breaks down into oxygen and water over a very, very long period of time.
   The experiment shown in figure 2.18 was set up to investigate the effect of catalase (an enzyme present in living cells) on the rate of this chemical reaction.

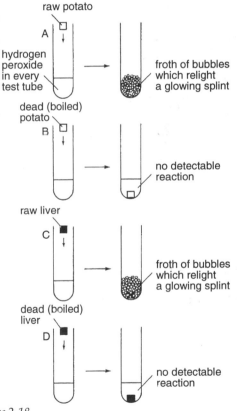

Figure 2.18

a) Identify the substrate in this experiment.
b) (i) Which TWO tubes received cells which provided a supply of active catalase?
(ii) Describe the effect that this enzyme had on the rate of breakdown of hydrogen peroxide.

(iii) Name ONE of the end products formed and describe how it was identified.

c) Which TWO tubes received denatured enzyme?

d) Explain at molecular level why a denatured enzyme is ineffective.

3 Figure 2.19 shows the stages that occur during an enzyme-controlled reaction.

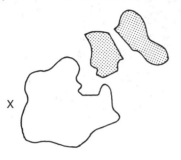

X

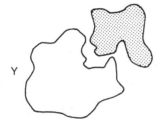

Y

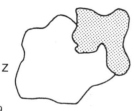

Z

Figure 2.19

a) What name is given to the complex indicated by the letter Z?

b) Using the THREE letters given, indicate the correct sequence in which the three stages occur if:

(i) the enzyme promotes the breakdown of a complex molecule to simpler ones;

(ii) the enzyme promotes the building-up of a complex molecule from simpler ones.

4 Molecules of simple sugar can be built up into starch molecules. The rate of this chemical reaction is speeded up by the enzyme phosphorylase.

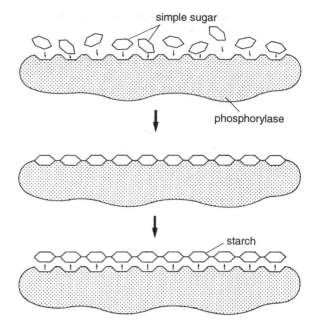

Figure 2.20

Figure 2.20 represents the action of phosphorylase at molecular level.

Describe the mechanism by which this enzyme is thought to work by using all of the following terms: active site on enzyme, enzyme, enzyme-substrate complex, lock-and-key, product, substrate.

5 Table 2.1 gives the results from an experiment set up to investigate the effect of temperature on the action of a digestive enzyme.

temperature (°C)	mass of substrate broken down (mg/h)
0	0
5	1
10	4
15	8
20	14
25	22
30	28
35	31
40	32
45	29
50	18
55	0

Table 2.1

a) Present the data as a line graph by plotting the points and joining them up with a curve.

b) (i) Explain what is meant by the term *optimum* temperature.

(ii) State the optimum temperature for the action of this enzyme.

c) (i) By how many times was the rate of enzyme activity greater at 30°C than at 20°C?

(ii) Explain the difference in terms of rate of molecular movement and frequency of collision between enzyme and substrate molecules at these two temperatures.

d) Which rise in temperature of 5°C brought about the biggest increase in rate of enzyme activity?

e) At the temperature range of 50–55°C, the molecules are still gaining energy so why does the reaction come to a halt at 55°C?

f) Predict the mass of substrate that would be broken down at 75°C.

6 Figure 2.21 shows an enzyme molecule at 35°C.

a) Name the type of protein of which this enzyme is composed.

b) Match the terms *weak hydrogen bond*, *active site* and *polypeptide chain* with the parts lettered X, Y and Z in the diagram.

c) Make a simple sketch of a molecule that could be this enzyme's substrate.

d) (i) Which lettered part of the diagram breaks when the temperature of the enzyme molecule is raised to 60°C.

(ii) What effect does this have on the arrangement of the polypeptide chains present in this protein molecule?

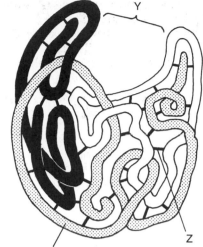

Figure 2.21 X

(iii) What effect does it have in turn on the enzyme's active site?

(iv) What word is used to describe an enzyme in this state?

7 The graph in figure 2.22 shows the effect of pH on the activity of three enzymes X, Y and Z.

a) State the working range of pH for each of the enzymes.

b) What generalisation can be made about:

(i) the breadth of working range of pH of each of the enzymes?

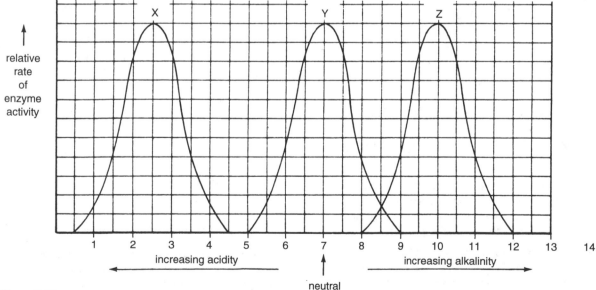

Figure 2.22

(ii) the extent to which they all share the same actual pH working range?

c) State the optimum pH for each of the enzymes.

d) Suggest which of the enzymes would show optimum activity in the human (i) mouth (ii) stomach.

e) Describe TWO adverse effects that a sudden change in pH can have on an enzyme's molecular structure.

8 One gram of roughly chopped raw liver was added to hydrogen peroxide solution at different pH values and the time taken to collect $1\,cm^3$ of oxygen was noted in each case. The results are given in table 2.2.

pH of hydrogen peroxide solution	time to collect $1\,cm^3$ of oxygen (seconds)
6	105
7	78
8	57
9	45
10	52
11	66
12	99

Table 2.2

a) Present the results as a line graph (in the form of a curve.)

b) From your graph state the pH at which the enzyme was (i) most active (ii) least active.

c) Of the pH values used in this experiment, which is the optimum for the enzyme present in the liver cells?

d) How could you obtain an even more accurate measurement of the optimum pH at which this enzyme works?

cylinder of hard boiled egg white in each tube

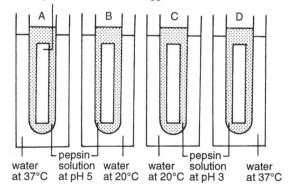

water at 37°C   pepsin solution at pH 5   water at 20°C   water at 20°C   pepsin solution at pH 3   water at 37°C

Figure 2.23

9 Pepsin is an enzyme which works best in strongly acidic conditions. It is secreted by glands in the lining of the human stomach.

a) In which of the tubes shown in figure 2.23 would most digestion of egg white occur?

b) Explain your choice of answer to **a)**.

10 Explain each of the following in terms of enzymes:

a) Fevers which raise the body temperature to over 42°C are normally fatal to human beings.

b) Vinegar (an acid) is used to preserve food against attack by micro-organisms.

c) Cheese kept in a warm room turns mouldy much more quickly than cheese kept in a fridge.

11 **a)** Draw a simple line graph to show how the rate of a typical enzyme-controlled reaction is affected by increasing enzyme concentration. (Assume that the substrate concentration is limited.)

b) Explain as fully as you can why the line graph takes the form that it does.

12 The graph shown in figure 2.25 summarises the results from an experiment involving an enzyme-controlled reaction.

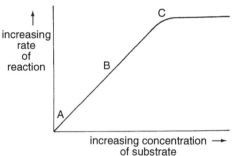

Figure 2.25

a) (i) In this experiment, the enzyme concentration was kept constant. From the graph, identify the factor that was varied by the experimenter.

(ii) What effect did an increase in this factor have over region AB of the graph.

b) Suggest which factor became limiting at point C on the graph.

c) Which letter on the graph represents the conditions in which

(i) almost all of the active sites (ii) none of the active sites (iii) a fair number of active sites on

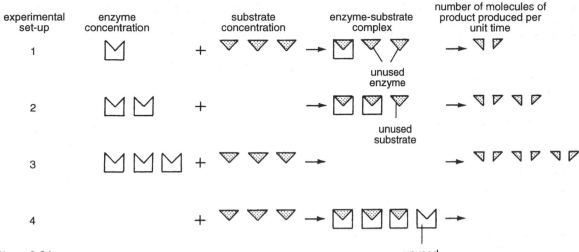

Figure 2.24

enzyme molecules were freely available for attachment to substrate molecules?
d) Suggest what could be done to increase the rate of the reaction beyond the level it reached at C.

13 Figure 2.24 represents, at molecular level, the effect of increasing enzyme concentration on the rate of an enzyme-controlled reaction.
a) Copy and complete the diagram to include the missing molecules.
b) From the information in the diagram, state which factor was (i) varied (ii) kept constant

during the series of experiments.
c) (i) Which piece of information in the diagram could be used as a measure of reaction rate?
(ii) What effect did increasing the enzyme concentration have on reaction rate in experimental set-ups 1, 2 and 3?
d) What factor was limiting in experimental set-up 4, keeping its reaction rate the same as in set-up 3?

14 Figure 2.26 represents, at molecular level, the effect of varying substrate concentration on the rate of an enzyme-controlled reaction.

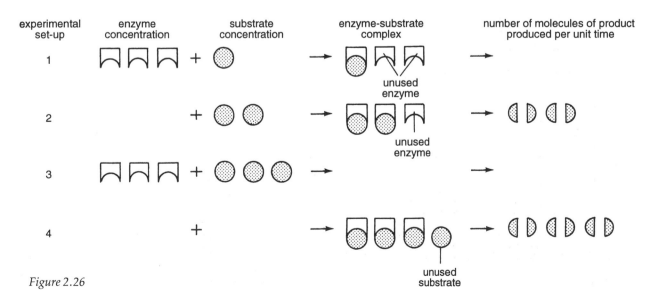

Figure 2.26

a) Copy and complete the diagram to include the missing molecules.
b) From the information in the diagram, state the factor that was kept constant by the experimenter.
c) (i) State the rate of the reaction in set-up 4.
(ii) Why was this rate equal to that in set-up 3 despite the presence of extra substrate in 4?
(iii) In which of the set-ups were active sites on enzyme molecules available since substrate molecules were in short supply?

15 Figure 2.27 refers to enzyme and inhibitor action at molecular level.
a) Which of the inhibitors is (i) competitive? (ii) non-competitive?
b) Justify your answer to a) by describing each inhibitor's mode of action.

c) Which type of inhibitor is (i) affected (ii) unaffected by the concentration of substrate present in the surrounding medium?

16 The flow diagram shown in figure 2.28 represents a complete biochemical pathway where substances P, Q and R are present in equal concentrations at the start.

$$\text{substance} \xrightarrow{\text{enzyme X}} \text{substance} \xrightarrow{\text{enzyme Y}} \text{substance}$$
$$\qquad P \qquad\qquad\qquad Q \qquad\qquad\qquad R$$

Figure 2.28

a) Name enzyme X's (i) substrate (ii) end product.
b) Name enzyme Y's (i) substrate (ii) end product.
c) Enzyme X is known to be affected by a certain inhibitor. Predict what will happen to

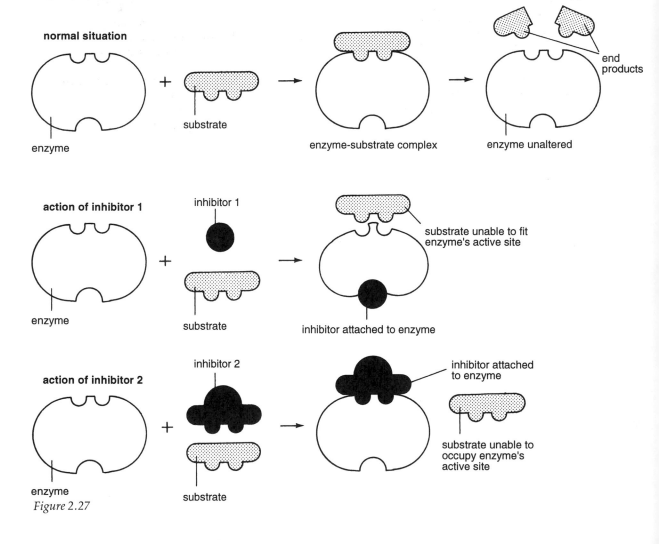

Figure 2.27

the concentrations of substances P, Q and R if a very high concentration of the inhibitor is added to the system.

d) Enzyme Y is known to be affected by a second inhibitor. Predict what will happen to the concentrations of substances P, Q and R if a very high concentration of this inhibitor is added to the system.

17 The graph in figure 2.29 summarises the results of three experiments using the same enzyme. In each case the concentration of the enzyme was kept constant.

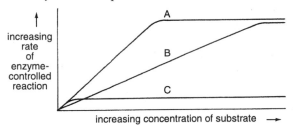

Figure 2.29

a) Which line graph shows the effect of increasing substrate concentration on the reaction where the enzyme is unaffected by an inhibitor?

b) Which line graph represents the effect of a non-competitive inhibitor bound to almost all of the enzyme molecules?

c) (i) Which line graph represents the effect of a competitive inhibitor?

(ii) Justify your choice of answer by explaining why this line graph takes the form that it does.

18 Rewrite the following sentences choosing the correct answer in each case.

Enzyme molecules are made of $\left\{\begin{array}{l}\text{fat}\\\text{protein}\\\text{carbohydrate}\end{array}\right\}$

and are produced by $\left\{\begin{array}{l}\text{all}\\\text{most}\\\text{a few}\end{array}\right\}$ living cells. Most

enzymes work best at around $\left\{\begin{array}{l}20°\text{C}\\40°\text{C}\\60°\text{C}\end{array}\right\}$ and are

described as biological $\left\{\begin{array}{l}\text{substrates.}\\\text{products.}\\\text{catalysts.}\end{array}\right\}$ In order

to work properly, many enzymes require the assistance of a $\left\{\begin{array}{l}\text{co-factor.}\\\text{inhibitor.}\\\text{competitor.}\end{array}\right\}$

3 EXPERIMENTS RELATING TO CELL PROCESSES

1 INVESTIGATING THE EFFECT OF TEMPERATURE ON AMYLASE ACTIVITY

INFORMATION

Amylase is an enzyme which promotes the breakdown of starch to simple sugar as shown in the following equation:

$$\text{starch} \xrightarrow{\text{amylase}} \text{simple sugar (maltose)}$$

Starch agar turns blue-black on being flooded with iodine solution. In the absence of starch, the agar fails to turn blue-black.

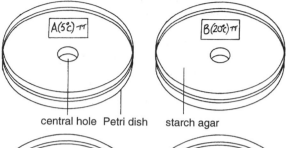
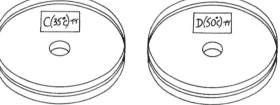

central hole Petri dish   starch agar

Figure 3.1: Apparatus before addition of amylase

Activity 1

a) Read all of the steps given in activities 1, 2 and 3 before carrying out the investigation.
b) Prepare for your investigation by collecting the following equipment:
    4 Petri dishes containing starch agar
    1 cork borer
    1 dropping bottle of amylase
    4 labels

Activity 2

Set up the experiment by carrying out the following instructions:
a) Label the 4 dishes A (5°C), B (20°C), C (35°C) and D (50°C) and add your initials to each label.
b) Using the cork borer, cut out a cylinder of starch agar to make a central hole in each dish.
c) Place the 4 unwanted cylinders of starch agar in the bin.
d) Prepare a table to record your results.
e) Ask your teacher to check that she/he is satisfied that you have carried out the above instructions as specified and have set up your equipment correctly. It should resemble figure 3.1.

Activity 3

Carry out the experimental procedure by doing the following:
a) Add an equal volume of amylase to almost fill the hole in each dish.
b) Place dish A in the fridge at 5°C.
c) Leave dish B out in the room at 20°C.
d) Place dish C in the incubator at 35°C.
e) Place dish D in the oven at 50°C.
f) After 24 hours collect the 4 dishes, a bottle of iodine solution and a paint brush.
g) Add an equal volume of iodine solution to each plate and spread it evenly over the surface with the paint brush.
h) The non blue-black zone round each enzyme-filled hole is the region where the enzyme has promoted the digestion of starch to simple sugar. Measure the diameter of this zone in dish A and subtract the diameter of the hole.
i) Repeat step h) for each dish and record the results in your table.
j) Ask your teacher to check that she/he is satisfied that your experimental procedure has been carried out correctly and safely.
k) Clear away your apparatus as instructed by your teacher.

Activity 4

Write up your report by doing the following:
a) State clearly the **aim** of the investigation.
b) Give a step-by-step account of the **procedure** that you followed both in preparing for and carrying out the experiment.
c) Make a final version of your table of **results**.
d) State any **conclusions** that you can draw from your results.
e) Think your experiment through critically and try to identify any **sources of error.** Say what these were and state the means by which they could be reduced in a repeat of the experiment. If there were no sources of error, say so and justify this answer by referring to one or more points of **good scientific practice** that you adopted during the investigation.
f) State how you could check the **reliability** of your results.

2 INVESTIGATING THE EFFECT OF pH ON PEPSIN ACTIVITY

INFORMATION

Pepsin is an enzyme found in the human stomach which promotes the breakdown (digestion) of insoluble protein molecules to soluble peptide molecules as in the following equation:

$$protein \xrightarrow{\text{pepsin}} peptides$$

Activity 1

a) Read all of the steps given in activities 1, 2 and 3 before beginning your investigation.

b) Prepare for your investigation by collecting the following equipment:
   4 glass tubes of equal length (e.g. 40 mm) containing boiled egg white
   4 glass beakers
   4 labels
   1 dropping bottle of pepsin
   4 dropping bottles of buffer solution (one each of pH 2, 4, 7 and 9)
   1 book of pH paper
   1 pair of plastic forceps
   1 paper towel
   1 pair of safety goggles
   1 ruler

Activity 2

Set up the experiment by carrying out the following instructions:
a) Label the 4 beakers A (pH 2), B (pH 4), C (pH 7) and D (pH 9) and add your initials to the labels.
b) Prepare a table to record your results and take a note of the initial length of solid egg white in each tube.
c) Wearing goggles, place a length of tubing containing egg into each beaker.
d) Just cover each with an equal volume of the appropriate buffer solution.
e) Using pH paper and plastic forceps, check that each beaker is at the required pH. (Remember to wipe the forceps with a paper towel between beakers and to take special care since acids are involved.)
f) Ask your teacher to check that she/he is satisfied that you have carried out the above instructions as specified and have set up your equipment correctly. It should resemble figure 3.2.

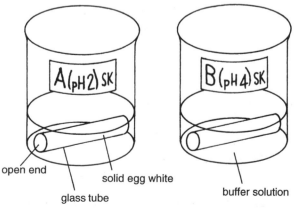
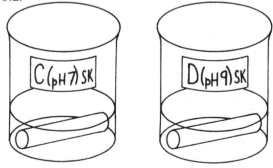

open end
solid egg white
glass tube
buffer solution

Figure 3.2: Apparatus before addition of pepsin

Activity 3

Carry out the experimental procedure by doing the following:

a) Ask your teacher how much pepsin solution to use and then add an equal volume to each beaker.

b) Check that each beaker is still at the required pH. Add more buffer if necessary and take care as before.

c) Place each beaker in the water bath at 37°C and leave for 24 hours.

d) After this time carefully remove the tubes of egg white using forceps and rinse them with cold water.

e) Measure the length of undigested egg white remaining in each tube.

f) Record the results in your table and calculate the length that has been digested.

g) Ask your teacher to check that she/he is satisfied that your experimental procedure has been carried out correctly and safely.

h) Clear away your apparatus as instructed by your teacher.

Activity 4

Write up your report by following the same instructions as those given in investigation 1 on page 33.

3 INVESTIGATING THE EFFECT OF LEAD AND CALCIUM ON CATALASE ACTIVITY

INFORMATION

Catalase is an enzyme which promotes the breakdown of hydrogen peroxide to water and oxygen as in the following equation:

$$\text{hydrogen peroxide} \xrightarrow{\text{catalase}} \text{water} + \text{oxygen}$$

Activity 1

a) Read all of the steps given in activities 1, 2 and 3 before beginning your investigation.

b) Prepare for your investigation by collecting the following equipment:
  1 test tube stand
  3 test tubes
  3 labels

1 dropping bottle of lead nitrate solution (1%)
1 dropping bottle of calcium nitrate solution (1%)
1 dropping bottle of distilled water
1 dropping bottle of hydrogen peroxide solution
supply of wooden splints
Bunsen burner
supply of finely chopped raw liver
1 pair of forceps
1 pair of safety goggles

Activity 2

Set up the experiment by carrying out the following instructions:

a) Label the test tubes A (water), B (lead) and C (calcium).

b) Place an equal amount of chopped liver in each test tube using the forceps.

c) Add drops of distilled water to tube A so that the liver is just immersed.

d) Add an equal volume of lead nitrate solution to tube B and calcium nitrate solution to tube C.

e) Allow the liver to soak in the chemicals for 5 minutes. During this time give the tubes an occasional shake and prepare a table for recording your results.

f) Light the Bunsen burner and position it carefully, ready for preparing glowing splints when required.

g) Ask your teacher to check that she/he is satisfied that you have carried out the above instructions as specified and have set up your equipment correctly. It should resemble figure 3.3.

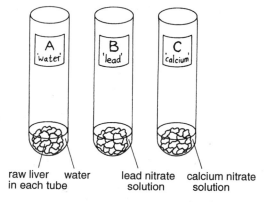

Figure 3.3:   Apparatus before addition of hydrogen peroxide

Activity 3

Carry out the experimental procedures by doing the following:

a) Put on the safety goggles.

b) Ask your teacher how many drops of hydrogen peroxide solution to use. (Take special care since hydrogen peroxide is a corrosive chemical. Any splashes on the skin must be washed off at once using cold water.)

c) Add the agreed number of drops of hydrogen peroxide to test tube A.

d) Immediately prepare a glowing splint and hold it at the mouth of test tube A.

e) Observe what happens both inside the test tube and at its mouth.

f) Repeat steps c), d) and e) for test tubes B and C.

g) Record the results in your table.

h) Ask your teacher to check that she/he is satisfied that your experimental procedure has been carried out correctly and safely.

i) Clear away your apparatus as instructed by your teacher.

Activity 4

Write up your report by following the same instructions as those given in investigation 1 on page 33.

SPECIMEN ANSWERS

INVESTIGATION 1

Aim

To investigate the effect of temperature on the activity of the enzyme amylase.

Procedure

1) 4 Petri dishes of starch agar were labelled A (5°C), B (20°C), C (35°C) and D (50°C) and initialled.

2) A central hole was cut out of the starch agar in each dish using a cork borer.

3) An equal volume of amylase was added to each central hole and then each dish was kept at its particular temperature for 24 hours.

4) After 24 hours, an equal volume of iodine solution was added to each plate and spread out over the surface using a paint brush.

5) The diameter of the non blue-black region round each enzyme-filled hole was measured and then the diameter of the hole was subtracted from each measurement.

6) The results were recorded as shown in table 3.1.

Results

dish	temperature (°C)	diameter of non blue-black zone (mm)
A	5	2
B	20	14
C	35	36
D	50	0

Table 3.1

Conclusions

As temperature increases, the activity of the enzyme amylase increases to an optimum at 35°C. At the higher temperature of 50°C, the enzyme is inactive.

Experimental errors

(i) The thickness of the starch agar was not exactly equal in the 4 dishes which meant that the hole in one dish was only about half-full of enzyme whereas in another the enzyme was almost overflowing. In a repeat of the experiment 4 dishes containing starch agar of equal thickness should be chosen.

(ii) Some dishes were left in light and some in darkness. This introduced a second variable factor which might have affected the result. In a repeat of the experiment, all the dishes should be covered with card to create uniformly dark conditions.

Reliability of results

This chould be checked by doing the experiment over again and finding out if the results could be repeated. If they could be repeated then this would show that the results were reliable.

INVESTIGATION 2

Aim

To investigate the effect of pH on the activity of the enzyme pepsin.

Procedure

1) 4 beakers were labelled A (pH 2), B (pH 4), C (pH 7) and D (pH 9) and initialled.

2) The initial length of egg white in each glass tube was noted and then one tube was put into each of the four beakers.
3) An equal volume of the correct buffer solution was added to each beaker and the pH was checked.
4) An equal volume of pepsin was added to each beaker and the pH in each checked again.
5) The 4 beakers were left at 37°C for 24 hours.
6) After 24 hours the final length of the egg white in each tube was measured and the length that had been digested was calculated.
7) The results were recorded as shown in table 3.2.

Results

pH	initial length of egg white (mm)	final length of egg white (mm)	length of egg white digested (mm)
2 (strongly acidic)	40	26	14
4 (acidic)	40	34	6
7 (neutral)	40	40	0
9 (alkaline)	40	40	0

Table 3.2

Conclusions

The enzyme pepsin works best at pH 2 (strongly acidic conditions). It works slightly at pH 4 (acidic conditions) but not at all at pH 7 (neutral) or pH 9 (alkaline).

Experimental errors

No sources of experimental error were identified in this investigation. All variable factors were keps constant except pH which was the one being investigated and good scientific practice was adopted throughout.

Reliability of results

This could be checked by doing the experiment over again and finding out if the results could be repeated. If they could be repeated then this would show that the results were reliable.

INVESTIGATION 3

Aim

To investigate the effect of lead and calcium on the activity of catalase.

Procedure

1) 3 test tubes were labelled A (water), B (lead) and C (calcium).
2) An equal amount of raw liver was placed in each test tube.
3) A volume of distilled water was added to A to just cover the liver.
4) An equal volume of lead nitrate solution was added to B and calcium nitrate solution to C.
5) The liver was allowed to soak in the chemicals for 5 minutes with occasional shaking.
6) After 5 minutes, drops of hydrogen peroxide solution were added to tube A.
7) A glowing splint was prepared and held at the mouth of test tube A.
8) Steps 6) and 7) were repeated for tubes B and C.
9) The results were recorded as shown in table 3.3.

Results

test tube	chemical present	state of contents of tube on adding hydrogen peroxide	effect on glowing splint at mouth of test tube
A	distilled water	violent reaction	relights
B	lead nitrate	no reaction	does not relight
C	calcium nitrate	violent reaction	relights

Table 3.3

Conclusion

The activity of the enzyme catalase is inhibited by lead but not by calcium.

Experimental error

It was difficult to add exactly the same amount of liver to each tube. In a repeat of the experiment, the liver should be cut up very finely and weighed out into three equal amounts in advance.

Reliability of results

This could be checked by doing the experiment over again and finding out if the results could be repeated. It they could be repeated then this would show that the results were reliable.

4 CHEMICAL PATHWAYS

ADENOSINE TRIPHOSPHATE (ATP)

A molecule of **ATP** is composed of adenosine and three inorganic phosphate (P_i) groups as shown in figure 4.1.

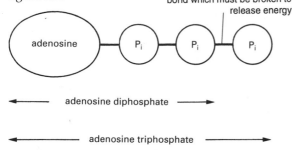

Figure 4.1: Structure of ATP

Energy stored in an ATP molecule is **released** when the bond attaching the terminal phosphate is broken by enzyme action. This results in the formation of adenosine diphosphate (ADP) and inorganic phosphate (P_i).

On the other hand, energy is required to regenerate ATP from ADP and inorganic phosphate by an enzyme-controlled process called **phosphorylation.** This reversible reaction is summarised by the equation:

$$\underset{\substack{\text{(high energy} \\ \text{state)}}}{\text{ATP}} \underset{\substack{\text{build-up} \\ \text{requiring energy}}}{\overset{\substack{\text{breakdown releasing energy}}}{\rightleftarrows}} \underset{\substack{\text{(low energy} \\ \text{state)}}}{\text{ADP} + P_i}$$

IMPORTANCE OF ATP

When an energy-rich substance such as glucose is broken down during respiration in a living cell, it releases energy which is used to make ATP. As a result, molecules of ATP are present in every living cell. Since ATP can rapidly revert to ADP + P_i, it is able to make energy become available for many energy-requiring processes (e.g. muscular contraction, protein synthesis and carbon fixation during photosynthesis) whenever it is needed.

CHEMISTRY OF RESPIRATION

Respiration is the process by which **chemical energy** is released during the breakdown (oxidation) of a foodstuff (e.g. glucose). It occurs in every living cell and involves the **regeneration of ATP** using energy from a complex series of biochemical reactions.

GLYCOLYSIS

In the cytoplasm of a living cell, the process of respiration begins with a molecule of **6-carbon glucose** being broken down by a series of enzyme-controlled steps to form two molecules of **3-carbon pyruvic acid** (see figure 4.2).

This process of 'glucose-splitting' is called **glycolysis.** It results in the net production of **2 ATP.** During glycolysis hydrogen is released from the respiratory substrate and becomes temporarily bound to a coenzyme molecule called **NAD.** (At no point in the pathway does hydrogen exist as free atoms or molecules.) A coenzyme molecule which has accepted hydrogen is said to be **reduced**.

The process of glycolysis does not require oxygen but the hydrogen bound to the reduced coenzyme (i.e. $NADH_2$) only produces further molecules of ATP (at a later stage in the process) if oxygen is present. In the absence of oxygen, **anaerobic respiration** occurs (see page 40).

MITOCHONDRIA

When oxygen is present, **aerobic respiration** occurs in the cell's **mitochondria** (see figure 4.3).

These are sausage-shaped organelles present in the cytoplasm of living cells. The inner membrane of each mitochondrion is folded into many plate-like extensions (**cristae**) which give a large surface area on which the processes of respiration can take place. The cristae protrude into the fluid-filled interior (**matrix**) which contains enzymes.

Each crista bears many stalked particles which are the site of ATP production. Cells requiring a lot of energy, such as sperm, liver, muscle and nerve cells, contain numerous mitochondria.

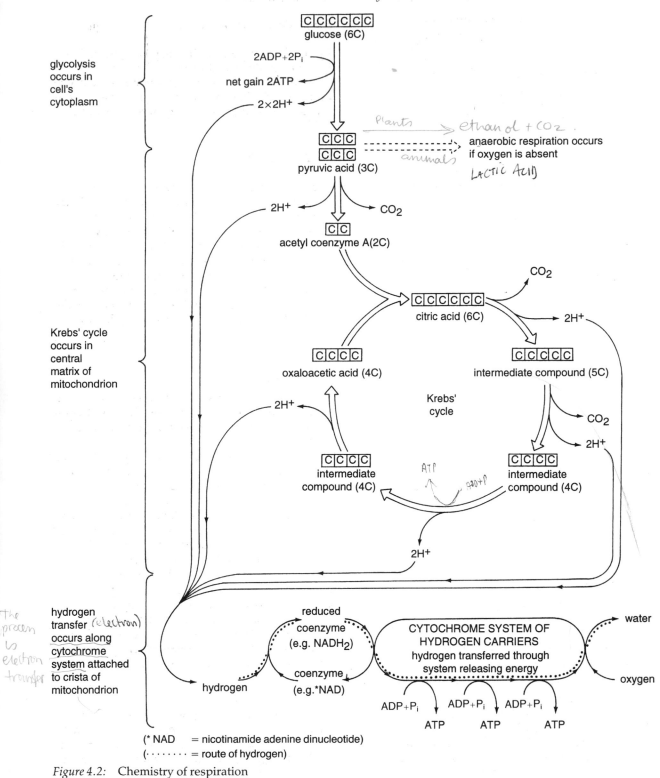

Figure 4.2: Chemistry of respiration

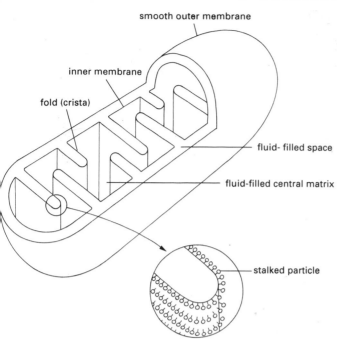

smooth outer membrane

inner membrane

fold (crista)

fluid- filled space

fluid-filled central matrix

stalked particle

Figure 4.3: Mitochondrion

FATE OF PYRUVIC ACID

Pyruvic acid produced during glycolysis diffuses into the matrix of a mitochondrion. Here it is converted into a **2-carbon** compound called **acetyl coenzyme A**. This reaction is accompanied by the release of hydrogen which again becomes bound to NAD.

KREBS' CYCLE

Each molecule of 2-carbon acetyl coenzyme A reacts with a molecule of **4-carbon oxaloacetic acid** present in the matrix of the mitochondrion to form **6-carbon citric acid.** This is gradually changed back to 4-carbon oxaloacetic acid via a 5-carbon and then several 4-carbon intermediate compounds. The cycle of enzyme-controlled reactions which brings about this conversion is also responsible for the release of **carbon dioxide** and **hydrogen** at various points along the way.

HYDROGEN TRANSFER

Figure 4.2 (which is a simplified version of the biochemical events taking place) shows that there are six points along the pathway where hydrogen is released and becomes temporarily bound to NAD.

These molecules of reduced coenzyme ($NADH_2$) transfer hydrogen to a chain of hydrogen carriers called the **cytochrome system.** Each mitochondrion has many of these systems attached to its cristae. Each time hydrogen passes along the cytochrome system, the energy released forms, on average, **3 molecules of ATP.** This process is called **oxidative phosphorylation.**

Figure 4.2 shows that the final hydrogen acceptor is **oxygen.** Hydrogen and oxygen combine (under the action of the enzyme cytochrome oxidase) to form **water.** Although oxygen only plays its part at the very end of the pathway, its presence is essential for hydrogen to pass along the cyotochrome system.

Total ATP

The aerobic part of the pathway results in the formation of 36 ATP from the breakdown of each glucose molecule. Since there is also a net gain of 2 ATP during glycolysis, the complete breakdown of one molecule yields a total of **38 ATP.**

KEY QUESTIONS

1 **a)** Give the full names of ADP and ATP.
b) Explain the difference between them in terms of energy state.
c) Why is it important that a living cell has a supply of ATP?

2 **a)** Into what substance is a molecule of glucose broken down during glycolysis?
b) How many molecules of ATP are gained by the cells as a result?
c) How many carbon atoms are present in a molecule of pyruvic acid?

3 **a)** Name the type of organelle responsible for aerobic respiration.
b) What is the central fluid-filled inner cavity of this organelle called?
c) What name is given to the folded extensions of this organelle's inner membrane which present a large surface area?

KEY QUESTIONS

1 a) Into what substance is pyruvic acid converted on entering a mitochondrion?
b) (i) What TWO substances are released during this breakdown?
(ii) Which of these becomes bound to a coenzyme molecule?

2 a) What substance is formed when acetyl coenzyme A reacts with oxaloacetic acid?
b) In which region of a mitochondrion does this reaction occur?
c) What name is given to the cycle of reactions by which citric acid is changed back to oxaloacetic?
d) What happens to the hydrogen released during the cycle?

3 a) What is the function of the cytochrome system?
b) Where is the cytochrome system located in a mitochondrion?
c) Name the high energy compound whose synthesis depends on energy released when hydrogen passes along the cytochrome system.

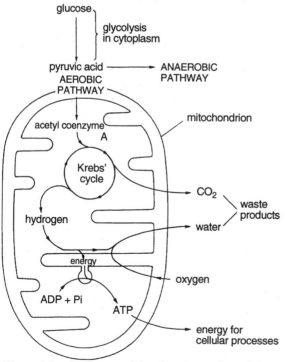

Figure 4.4: Summary of the chemistry of aerobic respiration

FUNCTION OF AEROBIC RESPIRATION

Aerobic respiration is summarised in figure 4.4. It is a metabolic pathway consisting of a series of enzyme-controlled reactions which bring about the oxidation of 6-carbon glucose.

The energy released is used to produce ATP from ADP + P_i. This **regeneration of ATP** for use in other cellular processes is the key function of repiration.

ANAEROBIC RESPIRATION

This is the process by which a little energy is derived from the **partial breakdown** of sugar in the absence of oxygen. Since oxygen is unavailable to the cell, the hydrogen transfer system and the Krebs' cycle (see figure 4.2) cannot operate in any of the cell's mitochondria. Only glycolysis can occur.

Each glucose molecule is partially broken down to pyruvic acid and yields only **two molecules of ATP.** The hydrogen released cannot go on to make ATP in the absence of oxygen. The pyruvic acid undergoes one of the following metabolic pathways (depending on the organism involved).

Anaerobic respiration in plants

The equation below summarises this process in plant cells such as yeast and water-logged roots:
glucose $\longrightarrow$ pyruvic acid $\longrightarrow$ ethanol + $2CO_2$
  (6C)   (2 × 3C)   (2 × 2C)

Anaerobic respiration in animals

The equation below summarises this process in animal cells such as skeletal muscle tissue:

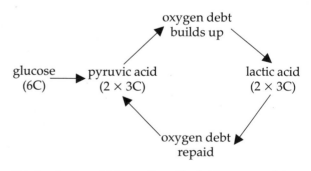

During lactic acid formation, the body accumulates an **oxygen debt.** This is repaid when oxygen becomes available and lactic acid is converted back

to pyruvic acid which then enters the aerobic pathway.

Anaerobic respiration is a less efficient process since it produces only **2 ATP** per molecule of glucose compared with **38 ATP** formed by aerobic respiration. The majority of living cells thrive in oxygen and respire aerobically. They only resort to anaerobic respiration to obtain a little energy for survival when oxygen is absent.

KEY QUESTIONS

1 **a)** With reference to oxygen, explain the difference between aerobic and anaerobic respiration.
b) State the number of ATP molecules formed from the breakdown of one glucose molecule during each type of respiration.

2 **a)** Give the word equation of anaerobic respiration in (i) a plant (ii) an animal cell.
b) Which of these forms of respiration is reversible? Explain your answer.

CHEMISTRY OF PHOTOSYNTHESIS

Photosynthesis is the process by which organic compounds are built up from molecules of carbon dioxide and water. The energy required for this process comes from **light**. The light energy is absorbed by photosynthetic pigments (e.g. green **chlorophyll**).

Photosynthesis consists of two separate parts: a light-dependent stage called **photolysis** and a temperature-dependent stage called **carbon fixation.** Photosynthesis occurs in chloroplasts.

CHLOROPLAST

Chloroplasts (see figure 4.5) are relatively large discus-shaped organelles situated in the cytoplasm of green plant cells. Each is bounded by a double membrane and possesses distinct internal structures made of **grana** and **lamellae.**

Each granum consists of a coin-like stack of flattened sacs containing molecules of **chlorophyll.**

Grana are the site of the light-dependent stage of photosynthesis. Lamellae are tubular extensions which form an interconnecting network between grana but do not contain chlorophyll.

The colourless background material in a chloroplast is called **stroma.** It is the site of the carbon fixation stage of photosynthesis. It lacks chlorophyll but contains important enzymes and starch grains (which act as temporary stores of photosynthetic products).

PHOTOLYSIS (LIGHT-DEPENDENT STAGE)

Solar (light) energy is trapped by chlorophyll in grana and converted into chemical energy. This process involves several important events which are summarised in figure 4.6 overleaf.

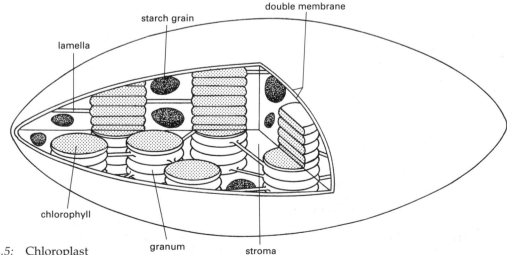

Figure 4.5: Chloroplast

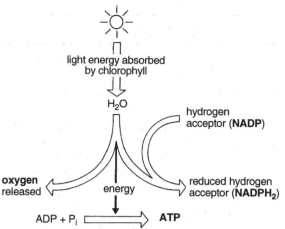

Figure 4.6: Photolysis (light-dependent stage)

Light energy is used to split molecules of water into hydrogen and oxygen. This is called the **photolysis of water.** The oxygen is released as a by-product. The hydrogen combines with a hydrogen acceptor called **NADP** (full name: nicotinamide adenine dinucleotide phosphate). The result is the formation of reduced hydrogen acceptor, $NADPH_2$.

In addition, chlorophyll makes energy available for the regeneration of ATP from ADP and inorganic phosphate. This process is called **photophosphorylation.**

The hydrogen held by $NADPH_2$ and the energy held by ATP at the end of the light stage are essential for use in carbon fixation, the second stage of photosynthesis.

KEY QUESTIONS

1 **a)** Define photosynthesis.
   b) (i) In which type of organelle does this process take place? (ii) Give ONE structural difference between the grana and the stroma in this organelle.

2 **a)** (i) Name the green pigment needed for photosynthesis.
   (ii) What type of energy is trapped by this pigment?
   (iii) What effect does this energy have on the water molecules during the light-dependent stage of photosynthesis?
   b) Describe the fate of the hydrogen and oxygen released as a result of photolysis.

3 **a)** What is meant by the term *photophosphorylation?*
   b) In which region of a chloroplast do photolysis and photophosphorylation occur?

CARBON FIXATION

This stage occurs in the stroma of a chloroplast. It consists of several enzyme-controlled chemical reactions which take the form of a cycle (often referred to as the **Calvin cycle** after the scientist who discovered it). Figure 4.7 summarises the cycle and indicates the number of carbon atoms in the molecules of the metabolites involved.

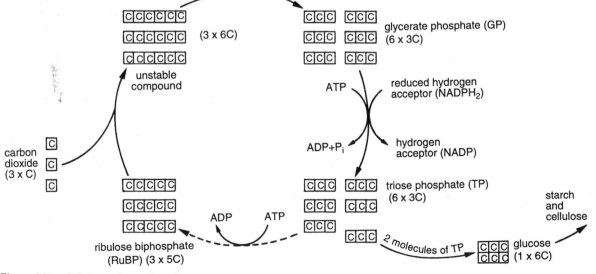

Figure 4.7: Calvin cycle (carbon fixation)

On entering a chloroplast by diffusion, a molecule of **carbon dioxide** combines with a molecule of **5-carbon ribulose biphosphate (RuBP)** to form a 6-carbon molecule. This molecule is unstable and rapidly splits into two molecules of **3-carbon glycerate phosphate (GP)**.

In the next stage of the cycle, GP is converted to a **3-carbon sugar** (**triose phosphate**). This conversion uses the hydrogen temporarily bound to $NADPH_2$ and some of the energy held in ATP. (Both of these are provided by the light-dependent stage.)

Each molecule of **glucose** (**a 6-carbon sugar**) is synthesised from a pair of these 3-carbon sugar molecules which combine together in an enzyme-controlled sequence of reactions. Molecules of glucose may then be built up into **starch** and **cellulose**.

However, the 3-carbon triose phosphate molecules are not all used to make complex products. Some are needed to regenerate RuBP, the carbon dioxide acceptor. This conversion of triose phosphate ($5 \times 3C$) to RuBP ($3 \times 5C$) also needs energy from ATP.

Summary

The process of photosynthesis involves the fixation of energy and is summarised in figure 4.8.

1 a) In which region of a chloroplast does the Calvin cycle occur?
b) Give another name for this stage of photosynthesis.

2 a) (i) Name the 5-carbon compound which acts as the carbon dioxide acceptor in the Calvin cycle. (ii) The 6-carbon compound formed is unstable. Name the 3-carbon compound into which it quickly breaks down.
b) Name the TWO substances from the light-dependent stage that must be present for glycerate phosphate to be converted to triose phosphate.

3 a) How many molecules of triose phosphate are needed to synthesise one molecule of glucose?
b) Most triose phosphate molecules remain in the cycle. Into what substance are they converted using ATP?

1 Match the terms in list **X** (which refer to the chemistry of respiration) with their descriptions in list **Y**.

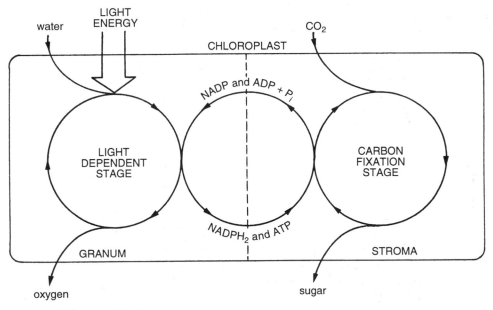

Figure 4.8: Summary of photosynthesis

list X

1) acetyl coenzyme A
2) ATP
3) carbon dioxide
4) central matrix
5) citric acid
6) crista
7) cytochrome system
8) cytoplasm
9) ethanol
10) glucose
11) glycolysis
12) Krebs' cycle
13) lactic acid
14) mito-chondrion
15) NAD
16) oxaloacetic acid
17) oxygen
18) phos-phorylation
19) pyruvic acid
20) water

list Y

a) stage of respiratory pathway common to both aerobic and anaerobic respiration

b) series of hydrogen carriers located on cristae of mitochondrion

c) stage of respiratory pathway that occurs in central matrix of mitochondrion

d) folded extension of inner membrane of mitochondrion

e) organelle responsible for aerobic respiration

f) the location in a cell where glycolysis occurs

g) part of a mitochondrion containing enzymes needed for Krebs' cycle

h) final hydrogen acceptor in aerobic respiration

i) product of aerobic respiration when oxygen combines with hydrogen

j) high energy compound formed by phosphorylation using energy released during respiration

k) coenzyme which accepts hydrogen during aerobic respiration and passes it to the cytochrome system

l) process by which high energy ATP is formed from low energy ADP + P_i

m) product of aerobic and anaerobic respiration in plant cells

n) 2-carbon compound formed from pyruvic acid in presence of oxygen

o) 2-carbon compound produced in plant cells during anaerobic respiration

p) 3-carbon compound formed from glucose during glycolysis

q) 3-carbon compound produced in animal cells during anaerobic respiration

r) 4-carbon compound in Krebs' cycle which combines with acetyl coenzyme A

s) 6-carbon compound formed when oxaloacetic acid and acetyl coenzyme A combine

t) 6-carbon sugar which acts as common respiratory substrate in both aerobic and anaerobic respiration

2 Figure 4.9 represents part of the summary of the chemistry of respiration as it occurs in a yeast cell.

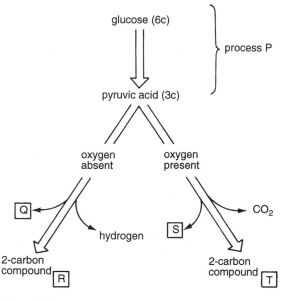

Figure 4.9

a) (i) Name process P.
(ii) Where in a yeast cell does process P occur?
(iii) How many molecules of ATP are gained by the yeast cell from the complete breakdown of one molecule of glucose?
b) What name is give to the type of respiration that occurs in (i) the absence (ii) the presence of oxygen? (iii) Which of these two forms of respiration generates less available energy for cellular work per molecule of glucose?
c) (i) Identify substances Q, R, S and T.
(ii) Predict the fate of substance Q.
(iii) State the effect of a high concentration of substance R on a yeast cell.
(iv) To what substance will S become combined?
(v) Into what substance will T be converted on meeting a molecule of oxaloacetic acid?

3 Figure 4.10 represents part of the chemical pathway that occurs in a cell during aerobic respiration.

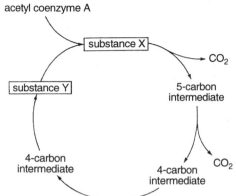

Figure 4.10

a) Identify substances X and Y.
b) What name is given to this series of reactions?
c) State the precise location of this process in the cell.
d) What substance is released at several stages in this cycle and immediately becomes bound to the coenzyme NAD?

4 Figure 4.11 shows a series of coupled reactions that occur along the hydrogen carrier system at the end of the aerobic respiration pathway.
a) (i) Name the final hydrogen carrier.
(ii) What form of respiration proceeds in the absence of this substance?
b) (i) By what other name is the hydrogen carrier system known?

(ii) Exactly where in a cell would this series of hydrogen carrier molecules be found?
c) (i) Give the full names of substances X, Y and Z.
(ii) Which of these contains energy which can readily be made available to the cell when required?

5 Figure 4.12 shows a simplified version of the chemistry of respiration in an animal cell.

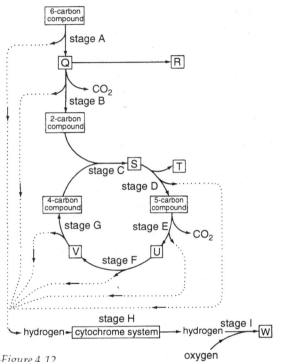

Figure 4.12

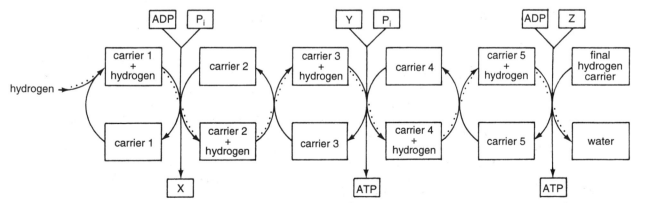

Figure 4.11   ($\cdots\cdots$ = route of hydrogen)

a) State the number of carbon atoms present in a molecule of each of the substances represented by boxes Q, R, S, T, V and Y.

b) Identify substances Q, R, S, T and W.

c) The various stages in the pathway are labelled using the letters A–I.

(i) At which of these stages is most ATP synthesised per molecule of glucose?

(ii) Identify another stage at which ATP is also synthesised (though in a much smaller quantity).

(iii) Why is the synthesis of ATP important to a cell?

(iv) Which of the lettered stages occurs in the cell's cytoplasm?

(v) Which stages would occur in the central matrix of a mitochondrion?

Exercises 6, 7, 8, 9 and 10 are multiple choice items and you should choose ONE correct answer in each case.

6 The enzymes required for the Krebs' cycle in a plant cell are located in the

A cytoplasmic fluid surrounding each mitochondrion.

B cristae of each mitochondrion.

C outer membrane of each mitochondrion.

D central matrix of each mitochondrion.

7 The final hydrogen acceptor in the cytochrome system is

A water.

B oxygen.

C NAD.

D ADP.

8 How many molecules of ATP are synthesised as a result of the complete oxidation of one molecule of glucose?

A 2

B 4

C 36

D 38

Questions 9 and 10 refer to the following possible answers.

A ethanol + CO_2 + ATP

B ethanol + ADP

C lactic acid + CO_2 + ADP

D lactic acid + ATP

9 Which answer correctly identifies the end products of anaerobic respiration in a water-logged root cell?

10 Which answer correctly identifies the end products of anaerobic respiration in mammalian muscle tissue?

11 Match the terms in list **X** (which refer to the first stage in the chemistry of photosynthesis) with their descriptions in list **Y**.

list X	list Y
1) ADP + P_i	**a)** product of photolysis of water which is required for aerobic respiration
2) chlorophyll	**b)** raw material which becomes split into oxygen and hydrogen during photosynthesis
3) granum	**c)** breakdown of water during light-dependent stage of photosynthesis
4) hydrogen	**d)** production of ATP using some of energy trapped during light-dependent reaction
5) light-dependent reaction	**e)** region of chloroplast containing molecules of chlorophyll
6) NADP	**f)** components of high energy compound formed by photo-phosphorylation
7) oxygen	**g)** compound which accepts hydrogen produced during photolysis of water
8) photolysis	**h)** green pigment which traps light energy
9) photophos-phorylation	**i)** product of photolysis of water which becomes attached to NADP
10) water	**j)** first stage in photosynthesis which involves photolysis of water

12 Match the terms in list **X** (which refer to the second stage in the chemistry of photosynthesis) with their descriptions in list **Y**.

List X	List Y
1) ATP	**a)** non green region of chloroplast containing enzymes
2) carbon dioxide	**b)** discus-shaped organelle responsible for photosynthesis
3) carbon fixation	**c)** reduced hydrogen acceptor needed for fixation of carbon in carbohydrate
4) chloroplast	**d)** 5-carbon compound which acts as the carbon dioxide acceptor

5) glucose **e)** 3-carbon sugar formed from glycerate phosphate

6) NADPH₂ **f)** first stable compound formed in Calvin cycle after carbon dioxide combines with its acceptor molecule

7) GP **g)** 6-carbon sugar molecule formed from two molecules of triose phosphate

8) RuBP **h)** raw material which supplies carbon atoms to be fixed into carbohydrate

9) stroma **i)** high energy compound which provides energy needed to drive the Calvin cycle

10) triose phosphate **j)** second stage in photosynthesis which is also known as the Calvin cycle

13 Figure 4.13 represents the light-dependent stage of photosynthesis.

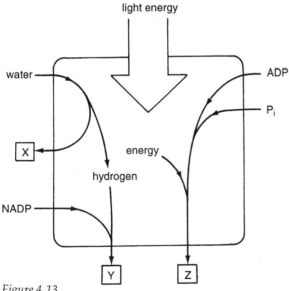

Figure 4.13

a) Give the colour and the name of the substance responsible for trapping light energy and making it available for use in the reaction shown in the diagram.
b) (i) Identify substances X, Y and Z.
(ii) Which of these is a by-product of the process and diffuses out of the cell?
(iii) Which of these go on to play important roles at a later stage in photosynthesis?

c) (i) In which region of a chloroplast does the light-dependent stage of photosynthesis occur?
(ii) State the energy conversion that takes place during this stage of photosynthesis.

14 Figure 4.14 shows the carbon fixation stage of photosynthesis.

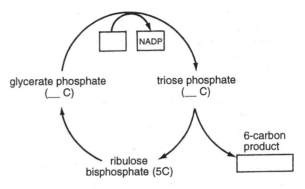

Figure 4.14

a) Copy the diagram and complete the blank boxes.
b) On your diagram, insert the number of carbon atoms possessed by a molecule of glycerate phosphate and triose phosphate.
c) (i) Which chemical acts as the carbon dioxide acceptor?
(ii) Add an arrow and the symbol CO_2 to your diagram to show where CO_2 enters the cycle.
d) (i) What is the full name of ATP?
(ii) Mark X on your diagram at TWO points at which ATP is needed for the reaction to proceed.
(iii) Why is ATP necessary at these points?
e) In which region of a chloroplast would carbon fixation be found to occur?

15 Figure 4.15 gives a simple outline of the chemistry of photosynthesis.

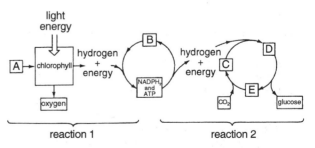

Figure 4.15

a) Give the names of the substances omitted from boxes A, B, C, D and E.
b) (i) Identify reactions 1 and 2.
(ii) Which of these reactions occurs in the grana of the chloroplast?
(iii) Where in a chloroplast does the other reaction occur?
c) Which of the two reactions is also referred to as the (i) Calvin cycle? (ii) photolysis of water?
d) Which products of reaction 1 are required to drive reaction 2?
e) During which reaction does photophosphorylation occur?
Exercises 16, 17, 18, 19 and 20 are multiple choice items and you should choose ONE correct answer in each case.

16 The first stable compounds resulting from the carbon fixation stage of photosynthesis are formed in the order
A glycerate phosphate ⟶ triose phosphate ⟶ glucose
B glucose ⟶ glycerate phosphate ⟶ triose phosphate
C glycerate phosphate ⟶ glucose ⟶ triose phosphate
D glucose ⟶ triose phosphate ⟶ glycerate phosphate

17 Photophosphorylation is the name given to the process by which
A chemical energy is converted into light energy in grana.
B ADP and inorganic phosphate are formed by the breakdown of ATP.
C light energy is absorbed by photosynthetic pigments in grana.
D ATP is synthesised during the light-dependent stage of photosynthesis.

Exercises 18, 19 and 20 refer to figure 4.16 of the cyclic series of reactions that occurs during the carbon fixation stage of photosynthesis.

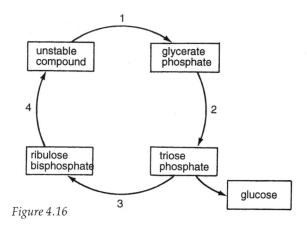

Figure 4.16

18 Carbon dioxide is taken into the cycle at stage
A 1.
B 2.
C 3.
D 4.

19 Hydrogen from reduced hydrogen acceptor is used at stage
A 1.
B 2.
C 3.
D 4.

20 Energy from ATP is used to drive stages
A 1 and 2.
B 2 and 3.
C 2 and 4.
D 3 and 4.

SECTION 2: Nucleic Acids and Protein Synthesis

5 MOLECULAR STRUCTURE OF NUCLEIC ACIDS

STRUCTURE OF DNA

Chromosomes are thread-like structures found inside the nucleus of a cell. They contain **deoxyribonucleic acid (DNA)**. A molecule of DNA consists of two strands each made of repeating units called **nucleotides.** Each DNA nucleotide (see figure 5.1) is made of a molecule of **deoxyribose sugar** joined to a **phosphate** group and a **base.** Since DNA posesses four different bases (**adenine, thymine, guanine** and **cytosine**) it has four different types of nucleotide.

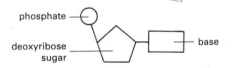

Figure 5.1: Structure of a DNA nucleotide

A strong **chemical bond** forms between the phosphate group of one nucleotide and the deoxyribose sugar of another. These bonds are not easily broken and join neighbouring nucleotide units into a permanent strand as shown in figure 5.2.

Two of these strands become joined together by weaker **hydrogen bonds** forming between their bases. However this union is temporary in that hydrogen bonds can be easily broken when this becomes necessary (e.g. during replication of DNA — see page 51).

Each base can only join with one other type: adenine (A) always bonds with thymine (T), and guanine (G) always bonds with cytosine (C). A—T and G—C are called **base pairs**.

The resultant double-stranded molecule is DNA and its two strands are arranged as shown in figure 5.3. This twisted coil is called a **double helix.** It is like a spiral ladder in which the sugar-phosphate 'backbones' form the uprights and the base pairs form the rungs.

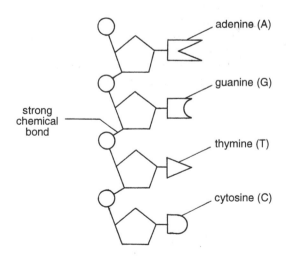

Figure 5.2: Strand of DNA nucleotides

49

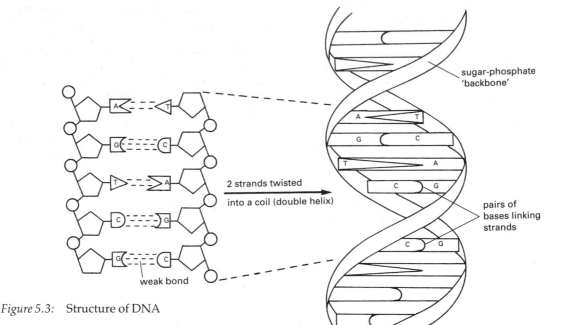

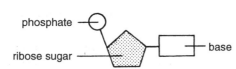

Figure 5.3:   Structure of DNA

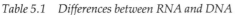

Figure 5.4:   Structure of an RNA nucleotide

STRUCTURE OF RNA

The second type of nucleic acid is called
ribonucleic acid (**RNA**). RNA also consists of
nucleotides (see figure 5.4). Although RNA's
structure (see figure 5.5) closely resembles that of
DNA, a molecule of RNA differs from a molecule
of DNA in three important ways as summarised in
table 5.1.

The functions of the two types of RNA
(messenger and transfer) will be discussed in
chapter 6.

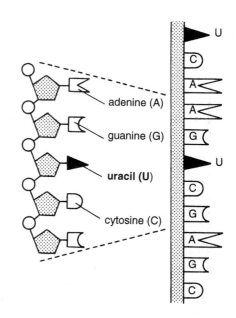

Figure 5.5:   Structure of RNA

	RNA	DNA
number of nucleotide strands present in one molecule	one	two
complementary base partner of adenine	uracil	thymine
sugar present in a nucleotide molecule	ribose	deoxy-ribose (each molecule contains one fewer oxygen atom than ribose)

Table 5.1   Differences between RNA and DNA

1 **a)** What is the full name of DNA?
b) What name is given to each of the repeating units that make up a nucleic acid strand?
c) How many strands are present in a molecule of DNA?

2 **a)** Name the THREE parts of a DNA nucleotide.
b) i) Which type of bond joins nucleotides into a strand of DNA? ii) Rewrite the following sentence and complete the blanks. This type of bond forms between the _____ molecule of one nucleotide and the _____ molecule of the next nucleotide on the nucleic acid strand.

3 **a)** How many different types of base molecule are found in DNA? Name each type.
b) Which type of bond forms between the bases of adjacent strands of a DNA molecule?
c) Describe the base-pairing rule.

4 **a)** What name is given to the twisted coil arrangement typical of a DNA molecule?
b) If DNA is like a spiral ladder, which part of it corresponds to the ladder's (i) rungs (ii) uprights?

5 **a)** Give the full name of RNA.
b) State THREE ways in which the structure of RNA differs from that of DNA.

REPLICATION OF DNA

DNA is a unique molecule because it is able to reproduce itself exactly. This process is called **replication** and it is the means by which new genetic material is made.

The process of replication is illustrated in figure 5.6. Stage 1 shows a region of the original DNA molecule after it has just become unwound. Stage 2 is a little further ahead of stage 1 in the process. Here weak hydrogen bonds between two bases are breaking and causing the two component strands of DNA to separate ('unzip') and expose their bases.

At stage 3, pairing of two bases enables a free DNA nucleotide to find its complementary nucleotide on the open chain. At stage 4, weak hydrogen bonds are forming between complementary base pairs.

At stage 5, slightly ahead of stage 4, a strong chemical bond is forming between the sugar of one nucleotide and the phosphate of the next one in the chain giving each strand its sugar-phosphate 'backbone'. This linking of nucleotides into a chain is called **polymerisation.** It is controlled by an enzyme called DNA polymerase.

Each stage 6 shows a newly formed daughter molecule of DNA about to wind up into a double helix. Daughter molecules have a base sequence identical to one another and to the original DNA molecule.

For DNA replication to occur the nucleus must contain:

- **DNA** (to act as a template for the new molecule);
- a supply of the four types of DNA **nucleotide**;
- the appropriate **enzymes** (e.g. DNA polymerase);
- a supply of **energy** from ATP (see page 37)

CONSERVATION OF BASE SEQUENCE

By conserving the base sequence throughout the process, DNA replication ensures that an **exact copy** of genetic information is passed from cell to cell during growth and from generation to generation during reproduction.

1 **a)** What is a molecule of DNA able to do that makes it unique compared to other chemical molecules?
b) What name is given to this process?

2 Study figure 5.6 carefully and then answer the following questions.
a) At which numbered stage has the DNA molecule been involved in the replication process for the shortest time?
b) (i) What type of bond is breaking at stage 2?
(ii) What effect does this have on the two component strands of the DNA molecule?
c) At which stage is base-pairing first seen to occur?
d) Name the type of bond formed and the types of molecule involved during (i) stage 4 (ii) stage 5.

Figure 5.6: Replication of DNA

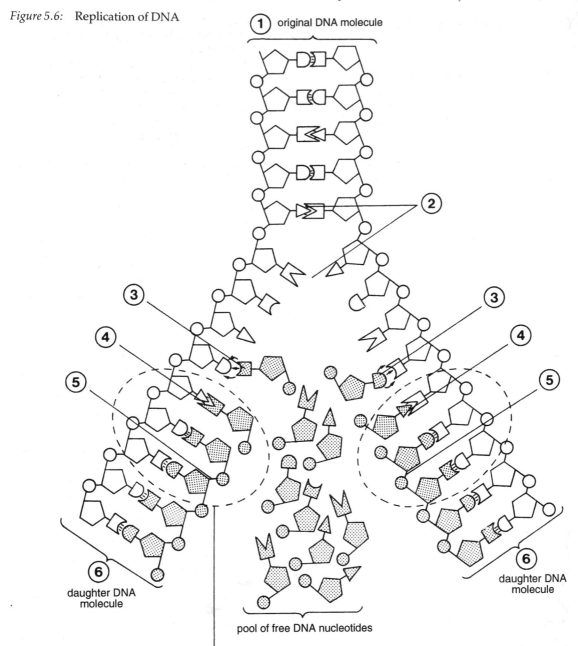

① original DNA molecule

②

③

④

⑤

⑥ daughter DNA molecule

⑥ daughter DNA molecule

pool of free DNA nucleotides

region of activity of enzyme (DNA polymerase)

3 What term is used to refer to the linking together of many similar molecules (such as DNA nucleotides) into a long chain?

4 Name FOUR substances that must be present in a nucleus for DNA replication to occur.

5 Why is it important that DNA molecule's particular base sequence is conserved from parent to daughter molecule?

TRANSCRIPTION OF mRNA FROM DNA

The genetic information carried on a section of DNA makes contact with structures in the cell's cytoplasm via a messenger. This go-between is called **messenger RNA (mRNA)** and it is formed (transcribed) from one of the DNA strands using free RNA nucleotides present in the cell's nucleus.

The process of **transcription** is illustrated in figure 5.7. Stage 1 shows the DNA strands becoming unwound. Stage 2 is a little further ahead of stage 1. Here weak hydrogen bonds between two bases are breaking and causing the DNA strands to separate.

At stage 3, pairing of bases enables a free RNA nucleotide to find its complementary nucleotide on the DNA strand which is being transcribed. At stage 4, weak hydrogen bonds are forming between two complementary bases.

At stage 5, a little further ahead in the process, a stronger chemical bond is forming between the sugar of one RNA nucleotide and the phosphate of the next one in the chain. This linking of nucleotides into a chain is called **polymerisation.** It is controlled by an enzyme called RNA polymerase.

At stage 6, the weak hydrogen bonds between the DNA and RNA bases are breaking allowing the molecule of transcribed mRNA to become separated from the DNA template.

Stage 7 shows transcribed mRNA ready to begin its journey out of the nucleus and into the cytoplasm.

At stage 8, weak hydrogen bonds between the two DNA strands reunite them and the molecule becomes wound up into a double helix once more.

Although replication of DNA and transcription of RNA have been presented as a series of stages, in reality each is a **continuous** process.

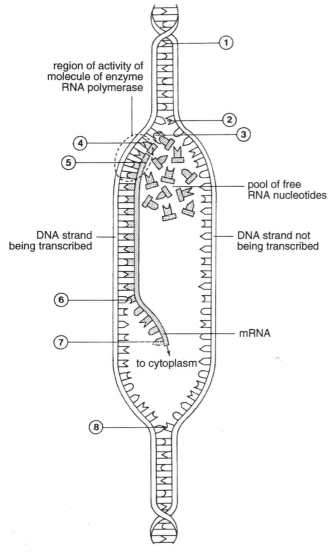

region of activity of molecule of enzyme RNA polymerase

pool of free RNA nucleotides

DNA strand being transcribed

DNA strand not being transcribed

mRNA

to cytoplasm

Figure 5.7: Transcription of mRNA

KEY QUESTIONS

1 **a)** By what means does DNA (which always stays in the nucleus) send messages to the cell's cytoplasm?
b) Name the messenger molecule.
c) What term is used to refer to the transfer of information from DNA to this messenger?

2 Study figure 5.7 carefully and then answer the following questions.
a) At which numbered stage has the DNA molecule been least involved in the transcription process so far?
b) (i) What type of bond is breaking at stage 2?
(ii) What effect does this have on the component strands of the DNA molecule?
c) At what stage is base-pairing first seen to occur?
d) Name the type of bond formed and the types of molecule involved during (i) stage 4
(ii) stage 5.

3 a) What term is used to refer to the linking together of many similar molecules (such as RNA nucleotides) into a long chain?
b) (i) What enzyme is needed to promote the linking of RNA nucleotides together into a long chain? (ii) What name is given to the molecule formed?

4 a) Name the type of bonds that break and the molecules involved during stage 6.
b) Why is this process important?

5 a) What happens to a molecule of transcribed mRNA?
b) Describe the behaviour of the DNA strands once transcription has been completed.

EXERCISES

1 Match the terms in list **X** with their descriptions in list **Y**.

list **X**	list **Y**
1) adenine	**a)** nucleic acid which is transcribed from a strand of DNA
2) chemical bond	**b)** basic unit of which nucleic acids are composed
3) chromo-some	**c)** enzyme which catalyses formation of sugar-phosphate bonds during DNA replication
4) cytosine	**d)** weak link between two base molecules which is easily broken
5) deoxyribose	**e)** complementary base partner of adenine in RNA
6) DNA	**f)** strong link between two molecules which is not easily broken
7) DNA polymerase	**g)** two-stranded molecule of DNA wound up into a spiral
8) double helix	**h)** complementary base partner of thymine in DNA
9) guanine	**i)** sugar present in DNA
10) hydrogen bond	**j)** process by which a complementary molecule of mRNA is made from a region of a DNA template
11) mRNA	**k)** complementary base partner of guanine in DNA and RNA
12) nucleotide	**l)** process by which a molecule of DNA reproduces itself

13) replication	**m)** enzyme which catalyses formation of sugar-phosphate bonds during transcription of mRNA
14) ribose	**n)** complementary base partner of adenine in DNA
15) RNA polymerase	**o)** nucleic acid present in chromosomes
16) thymine	**p)** sugar present in RNA
17) trans-cription	**q)** threadlike structure found inside the nucleus of a cell
18) uracil	**r)** complementary base partner of cytosine in DNA and RNA

Exercises 2–7 are multiple choice items. You should choose ONE correct answer only in each case.

2 The structure of one nucleotide is shown in figure 5.8.

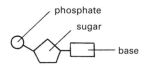

Figure 5.8

Which of diagrams in figure 5.9 shows two nucleotides correctly joined together?

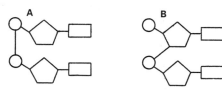

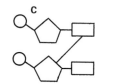

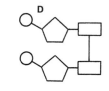

Figure 5.9

3 Which of the following is a base pair normally present in DNA?
A adenine and cytosine
B guanine and adenine
C thymine and guanine
D thymine and adenine

4 Which of the following is correct?

	present in DNA	present in RNA
A	uracil	thymine
B	deoxyribose	ribose
C	single strand	double strand
D	4 different nucleotides	5 different nucleotides

5 One of the nucleotides present in mRNA has the composition
  A adenine—ribose—phosphate
  B uracil—deoxyribose—phosphate
  C thymine—ribose—phosphate.
  D guanine—deoxyribose—phosphate.

6 A shorthand method of representing part of a single strand of DNA is shown in figure 5.10.

Figure 5.10

Which of the diagrams in figure 5.11 shows the correct positions of the phosphate (P), sugar (S) and base (B) molecules?

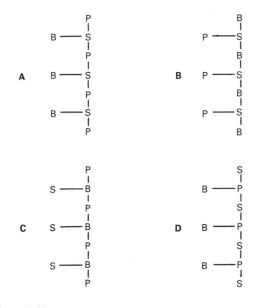

Figure 5.11

7 Strand X in figure 5.12 shows a small part of a nucleic acid molecule.

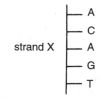

Figure 5.12

Which two strands shown in figure 5.13 are complementary to strand X?

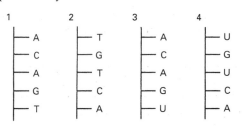

Figure 5.13

  A 1 and 3
  B 2 and 4
  C 1 and 2
  D 3 and 4

8 Figure 5.14 shows the molecular structure of two of the chemical units from which DNA is built.

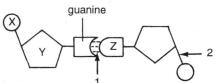

Figure 5.14

  a) Name this type of basic unit from which nucleic acids are constructed.
  b) Identify chemical molecules X, Y and Z.
  c) Redraw the diagram and then expand it to include the bases adenine and thymine connected to the appropriate molecules.
  d) Which of the numbered arrows indicates:
  (i) strong chemical bonding?
  (ii) weak hydrogen bonding?
  e) Name a structural feature of an inactive DNA molecule which the diagram above fails to illustrate.

9 Figure 5.15 represents a small portion of a molecule of RNA.
  a) Identify molecules 1, 4 and 5.

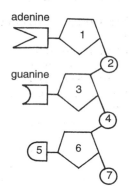

Figure 5.15

b) Name the fourth type of base molecule not included in the diagram.
c) How many nucleotides have been drawn in this diagram?
d) Using numbers only, represent one complete nucleotide.

10 Decide whether each of the following statements is true or false and then use T or F to indicate your choice. Where a statement is false, give the word(s) that should have been used in place of the word(s) in **bold print**.
a) Messenger RNA is a **protein.**
b) When inactive, DNA takes the form of a **double helix.**
c) The bases guanine, cytosine, adenine and **thymine** are present in mRNA.
d) Transfer of information from DNA to mRNA is called **replication.**
e) Formation of sugar-phosphate bonds is controlled by a **polymerase** enzyme.

11 A certain DNA molecule is found to contain 2000 bases of which 480 are guanine.
a) Calculate the number of thymine molecules present in this DNA molecule.
b) Express the number of guanine molecules as a percentage of the total number of bases present in the DNA molecule.

12 A DNA molecule is found to contain 10 000 base molecules of which 20 per cent are thymine.
a) What percentage of the bases are cytosine?
b) How many of the bases are adenine?

13 Place the following events that occur during DNA replication in the correct order.
A Bonds between opposite bases break.
B Bonds between opposite bases form.
C DNA molecule uncoils from one end.

D Opposite strands of DNA separate.
E Daughter DNA molecules coil into double helices.
F Sugar-phosphate bonds form.

14 Figure 5.16 shows a small part of a DNA molecule during the process of replication. (A = adenine and G = guanine.)

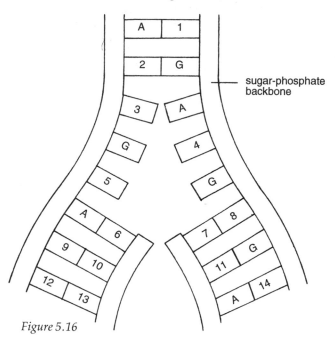

Figure 5.16

Identify bases 1–14.

15 Name FOUR substances that must be present in a nucleus for transcription of mRNA from DNA to occur.

16 Figure 5.17 shows a molecule of mRNA being transcribed from DNA.

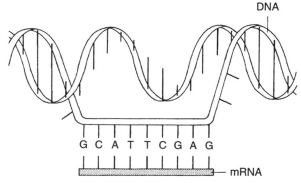

Figure 5.17

Redraw the mRNA to show the sequence of its bases.

17 The stages in the following list refer to the process of transcripion of mRNA from DNA.
   A Bases along each DNA strand become exposed.
   B Adjacent RNA nucleotides become joined to one another to form mRNA.
   C DNA strands become separated as hydrogen bonds break between base pairs.
   D Completed molecule of mRNA becomes detached from DNA template.
   E A specific region of a DNA molecule unwinds.
   F Hydrogen bonds form between bases of DNA strands which then rewind.
   G Each base on the DNA strand attracts its complementary base on a free RNA nucleotide.

   a) Arrange the stages into the correct sequence beginning with E.
   b) (i) Which stage describes the process of polymerisation?
   (ii) Name the enzyme which is necessary for this process to occur.

6 GENETIC CODE AND PROTEIN SYNTHESIS

SEQUENCE OF DNA BASES

When one living organism (e.g. a human) is compared with another very different living organism (e.g. a fruit fly), their cells are found to contain the same basic chemical compounds.

The chemical components of DNA also remain constant from species to species. However the DNA of one species differs from that of another in quantity and in the order in which the bases (A, T, G and C) occur along its length. It is this **sequence of bases** along the DNA strands which is unique to the organism. It contains the **genetic instructions** which control the organism's inherited characteristics.

PROTEIN STRUCTURE

Each molecule of protein is built up from a large number of sub-units called **amino acids** (of which there are about twenty different types). These are joined together by **peptide bonds** to form one or more very long chains called **polypeptides.** (See page 2).

ENZYMES

Every **enzyme** is made of protein. Its exact molecular structure, shape and ability to carry out its function all depend on the **sequence** of its amino acids.

An organism's inherited characteristics are the result of many biochemical processes controlled by enzymes. In humans, for example, certain enzymes govern the biochemical pathways which lead to the formation of hair of a certain type. In fruit flies different enzymes control the formation of bristles of a certain type.

An organism's DNA determines such inherited characteristics typical of the species by controlling the structure of the necessary enzymes. Each enzyme is constructed with its amino acids arranged in the particular order needed to perform its function. This critical order is determined by the sequence of the bases in the organism's DNA.

GENETIC CODE

The information present in DNA takes the form of a molecular code language called the **genetic code.** The sequence of bases along a DNA strand represents a sequence of 'codewords'.

DNA possesses only four different bases yet proteins contain about twenty different types of amino acid. The relationship cannot be one base coding for one amino acid since this would only allow four amino acids to be coded. Even two bases per amino acid would give only sixteen (4^2) different 'codewords'.

CODON

However if the bases are taken in groups of three then this gives 64 (4^3) different combinations (see Appendix 1). It is now known that each amino acid is coded for by one (or more) of these 64 **triplets** of bases. Each triplet is called a **codon.** The codon is the basic unit of the genetic code.

Thus a species' genetic information is encoded in its DNA with each strand bearing a series of base triplets arranged in a **specific order** for coding the particular proteins needed by that species.

KEY QUESTIONS

1 a) What are the THREE different components that make up each nucleotide in a DNA molecule? (see page 49 for help.)
b) What can be said about the chemical components of DNA with respect to all the forms of life that possess it?
c) In what way does the DNA of one species differ from that of another which makes each unique?

2 a) Name the sub-units of which all proteins are composed.
b) How many different types of these sub-units are found to occur in proteins?
c) By what means are these sub-units joined together into a chain?
d) Give the name of such a chain and make a simple diagram of one. (See page 2 for help.)

3 a) What is an enzyme?
b) Upon what do an enzyme's structure, shape and ability to carry out its function all depend?
c) What in turn determines the all-important sequence of amino acids in a protein?

4 a) With reference to the relationship between the genetic code and the protein synthesised, why is it not possible that
(i) one base corresponds to one amino acid?
(ii) two bases correspond to one amino acid?
b) How many bases in the genetic code do correspond to one amino acid in a polypeptide?
c) What name is given to these groups of bases which make up the basic units of the genetic code?

PROTEIN SYNTHESIS

mRNA

The region of DNA which carries the genetic code for the protein to be synthesised, temporarily splits open to expose its bases.

A molecule of mRNA is transcribed from one of the DNA strands as described on page 53. The completed mRNA molecule leaves the nucleus and enters the cytoplasm (see figure 6.1). Each triplet of bases on mRNA is called a **codon.**

tRNA

A second type of RNA is found in the cell's cytoplasm. This is called **transfer RNA (tRNA)** (see figure 6.1.). Each molecule of tRNA has one of its triplets of bases exposed. This triplet, known as

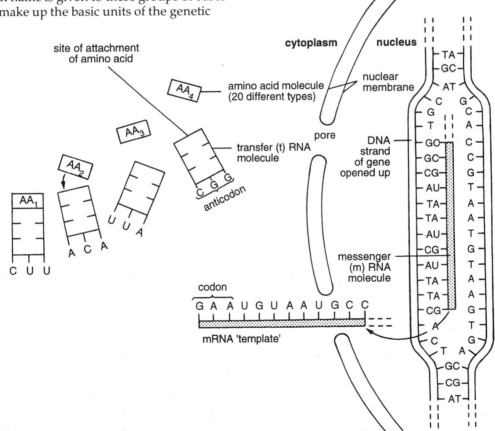

Figure 6.1:   Two types of RNA

an **anticodon**, corresponds to a **particular** amino acid. Each tRNA molecule picks up the appropriate amino acid from the cytoplasm at its **site of attachment.** Every cell has at least twenty different types of tRNA, one for each type of amino acid.

KEY QUESTIONS

1 a) By what means is a molecule of mRNA formed?

b) Where in a cell is the site of mRNA formation?

c) Where in a cell does mRNA carry out its function?

d) What name is given to each triplet of bases on mRNA?

2 a) Name a second type of RNA found in a cell's cytoplasm.

b) What name is given to a triplet of exposed bases on a tRNA molecule?

c) To what type of chemical molecule does each tRNA triplet correspond?

d) Name the region of a tRNA molecule to which its amino acid becomes joined.

e) Approximately how many types of tRNA occur in a normal living cell?

RIBOSOMES

These are small, almost spherical structures found in the cytoplasm of all living cells. They are the site of **translation** of RNA into protein. Each ribosome contains enzymes which are essential for the process of protein formation.

TRANSLATION OF RNA INTO PROTEIN

A ribosome becomes attached to one end of the mRNA molecule about to be translated. Inside the ribosome there are sites for the attachment of tRNA molecules, two at a time (see figure 6.2).

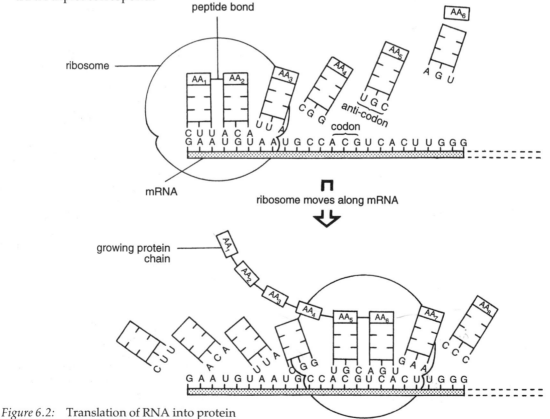

Figure 6.2:   Translation of RNA into protein

This arrangement allows the **anticodon** of the first **tRNA** molecule to form weak hydrogen bonds with the complementary **codon** on the **mRNA**. When the second tRNA molecule repeats this process, the first two amino acid molecules are brought into line with one another. They quickly become joined together by a strong peptide bond whose formation is controlled by an enzyme present in the ribosome.

The first tRNA becomes disconnected from its amino acid and from mRNA and then leaves the ribosome. The ribosome moves along the mRNA strand allowing the anticodon of the third tRNA to move into place and link with its complementary codon on mRNA. This allows the third amino acid to become bonded to the second one and so on along the chain.

Thus the process of translation brings about the alignment of amino acids in a certain order at the ribosome where they undergo **polymerisation** into a **polypeptide** chain. The completed polypeptide (consisting of very many amino acids) is then released into the cytoplasm.

Formation of completed protein may involve folding or rearrangement of the polypeptide chain (see page 2). Sometimes a number of polypeptide chains combine to form a protein molecule.

Each tRNA molecule becomes attached to another molecule of its amino acid ready to repeat the process. The mRNA is often reused to produce further molecules of the same polypeptide.

KEY QUESTIONS

1 a) What is a ribosome?
b) (i) What name is given to the conversion of the genetic message held by RNA into protein?
(ii) At what point on a molecule of mRNA does a ribosome become attached at the start of this process?

2 a) What type of bonds form between mRNA codons and tRNA anticodons brought together at a ribosome?
b) What type of bond forms between adjacent amino acids brought together by their tRNAs at a ribosome?
c) What type of substance controls the formation of bonds between amino acids?
d) What happens to a tRNA molecule once its amino acid has become bonded to another amino acid at a ribosome?

3 What is formed as a ribosome continues to move along a mRNA molecule?

4 What term is given to refer to the linking together of many similar molecules (such as amino acids) into a long chain?

5 Is it valid to describe mRNA and tRNA as *reusable* molecules? Justify your answer.

EXERCISES

1 Match the terms in list **X** with their descriptions in list **Y**.

list X	list Y
1) amino acid	a) sub-cellular structure which is the site of protein synthesis
2) anticodon	b) a type of protein molecule which controls the rate of a biochemical reaction
3) codon	c) type of nucleic acid which conveys information from DNA to a ribosome
4) enzyme	d) triplet of bases on a tRNA molecule which is complementary to a mRNA codon
5) genetic code	e) long chain of amino acids formed at a ribosome during translation of RNA
6) mRNA	f) the conversion of the genetic code into a sequence of amino acids in a polypeptide
7) polypeptide	g) molecular language made up of 64 codewords called codons
8) ribosome	h) unit of genetic code consisting of three mRNA bases
9) translation	i) type of nucleic acid which acts as an amino acid carrier
10) tRNA	j) one of 20 different types of sub-unit which make up protein molecules

2 Choose the ONE correct answer to each of the following questions, all of which relate directly or indirectly to the genetic code.
a) How many different types of amino acid are found to occur in proteins?
A 3   B 20   C 30   D 64
b) If each base in DNA coded for one type of amino acid only, how many types of amino acid would fail to have a code?
A 2   B 16   C 20   D 64

c) How many different codes can be formed by taking DNA's bases two at a time?
A 2   **B** 16   **C** 20   **D** 64

d) If each type of amino acid corresponded to a code of two bases, how many types of amino acid would lack a code?
A 2   **B** 4   **C** 16   **D** 20

e) How many DNA bases are now known to code for a molecule of amino acid?
A 1   **B** 2   **C** 3   **D** 4

f) How many different codons does the arrangement referred to in question e) allow?
A 4   **B** 16   **C** 20   **D** 64

g) How many base molecules are present in one codon?
A 1   **B** 2   **C** 3   **D** 4

3 The unique nature of an organism's DNA molecules arises from the sequence of its
A phosphates.
B bases.
C sugars.
D amino acids.
(Choose ONE correct answer only.)

4 Copy and complete table 6.1 by inserting the name(s) of the appropriate nucleic acids.

characteristic or function	nucleic acid(s)
contains the base uracil possesses anticodons composed of a double strand acts as an amino acid carrier contains the base thymine possesses codons composed of a single strand	

Table 6.1

5 a) Name the cellular organelle which is the site of translation of nucleic acids to polypeptides in a cell.
b) Where in a cell are these sub-cellular structures found?

6 Decide whether each of the following statements is true or false and then use T or F to indicate your choice. Where a statement is false, give the word(s) that should have been used in place of the word(s) in **bold print**.
a) The process by which information present in the base sequence of mRNA is used to produce a sequence of amino acids in a protein is called **transcription**.
b) A triplet of nucleotide bases present in

mRNA that codes for a specific amino acid during protein synthesis is called a **codon**.
c) A single-stranded molecule of nucleic acid which carries a transcribed version of the genetic code to the sites of protein synthesis is called **transfer RNA**.
d) The site of protein synthesis in a cell is the **nucleus**.
e) The process by which a large molecule (e.g. polypeptide) is built up by many simpler units (e.g. amino acids) bonding together is called **polymerisation**.

Exercises 7–10 are multiple choice items. In each case you should choose ONE correct answer only.

7 A mRNA template is
A translated from protein.
B transcribed into protein.
C translated into DNA.
D transcribed from DNA.

8 A free transfer RNA molecule can combine with
A one specific amino acid only.
B any available amino acid.
C three different amino acids.
D a chain of amino acids.

9 If each amino acid molecule weighs 100 mass units, what is the weight (in mass units) of the protein molecule synthesised from a mRNA molecule which is 600 bases long?
A 2000   **B** 6000   **C** 20000   **D** 60000

10 Table 6.2 shows three different mRNA molecules (each containing a base sequence) and the three different protein molecules synthesised from them.

mRNA	repeating sequence	protein
AGAGAGAGAGAGAGAGAG – – – –	AG	X
CAUCAUCAUCAUCAUCAU – – – –	CAU	Y
AAUGAAUGAAUGAAUGAAUGAAUG –	AAUG	Z

Table 6.2

Which of the answers in table 6.3 shows the correct number of different types of amino acid in each protein molecule?

	X	Y	Z
A	2	1	4
B	1	3	2
C	2	1	3
D	3	1	4

Table 6.3

11 Rewrite the following passage and complete the blanks.

DNA, situated in the _____ of a cell, controls protein synthesis by passing on the _____ information during the formation of a second nucleic acid called _____ . This molecule then passes out of the nucleus through a pore into the _____ where it becomes associated with sub-cellular structures called _____. These organelles supply some of the _____ needed for protein synthesis. It is here that tRNA molecules, each with an attached molecule of _____ _____ , line up on the surface of the mRNA molecule with codons matching complementary _____ . Adjacent amino acid molecules form _____ bonds and become joined into a _____ chain.

12 Figure 6.3 shows the method by which the genetic code is transmitted during protein synthesis.

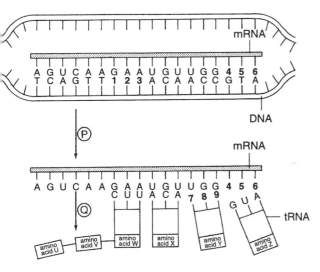

Figure 6.3

a) Identify bases 1–9.
b) What name is given to a triplet of bases on a molecule of (i) mRNA (ii) tRNA?
c) Name processes P and Q.
d) With reference to table 6.4, identify amino acids U, V, W, X, Y and Z.

codon	amino acid
GCG	alanine
UGU	cysteine
CAA	glutamine
GAA	glutamic acid
CAU	histidine
CCC	proline
AGU	serine
UAU	tyrosine
UGG	tryptophan

Table 6.4

13 a) With further reference to table 6.4 given in exercise 12, work out the mRNA code for part of a polypeptide chain with the amino acid sequence:

tyrosine—alanine—proline

b) State the genetic code on DNA from which this mRNA would be transcribed.

14 Figure 6.4 shows a strand of mRNA which carries the genetic code for a peptide consisting of fifteen amino acids.

Analysis of the peptide gave the results shown in table 6.5.

name of amino acid	abbreviated name	number of molecules of amino acid present in peptide
alanine	ala	5
arginine	arg	2
glycine	gly	1
lysine	lys	3
valine	val	4

Table 6.5

Using the abbreviated names, make a simple diagram to show the sequence in which the amino acids occur along the peptide chain.

A AG G U U GC C G C C C G G A A GG U U G CC A A G C G G G C C G G C G C C G U U G U U

↑
Code starts
here

Figure 6.4

15 Use the information in table 6.6 to help you answer a) and b) below.

amino acid	base anticodon (tRNA)
asparagine	UUA
glutamic acid	CUU
proline	GGA
threonine	UGG
tyrosine	AUA

Table 6.6

a) Give the base codon (mRNA) for each amino acid.

b) Draw a diagram of the sequence of amino acids that would be formed from the portion of mRNA shown in figure 6.5.

A A U G A A U A U G A A C C U A A U A C C

↑ start here

Figure 6.5

Exercises 16, 17 and 18 are multiple choice items. In each case you should choose ONE correct answer only. The three items all refer to figure 6.6 which shows the synthesis of part of a protein molecule.

16 Which of the following is the first part of the protein molecule that would be translated from mRNA₂?

start of
protein

↓

A $AA_4 - AA_2 - AA_7 - AA_6 - \ldots$
B $AA_6 - AA_7 - AA_2 - AA_4 - \ldots$
C $AA_3 - AA_1 - AA_5 - AA_8 - \ldots$
D $AA_8 - AA_5 - AA_1 - AA_3 - \ldots$

17 The DNA strand from which mRNA₂ was synthesised is
A GAACTGGACCCT
B CTTGACCTGGGA
C GAACUGGACCCU
D CUUGACCUGGGA

18 Figure 6.7 shows a small part of a different protein that was also synthesised on this ribosome.

Figure 6.7

What sequence of bases in DNA coded for this sequence of amino acids?
A CAGGUCAAGUCC
B CAGGTCAAGTCC
C GUCCAGUUCAGG
D GTCCAGTTCAGG

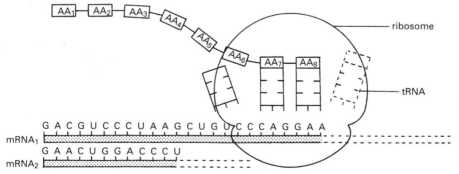

Figure 6.6

7 GENE EXPRESSION AND GENE MUTATION

GENE

A **gene** is a unit of heredity which controls a genetically inherited characteristic (e.g. hair colour, insulin production etc.). Each gene occupies a specific site on a chromosome. Each chromosome is a thread-like structure containing DNA (see figure 7.1).

A gene is a **segment of DNA.** Genes vary in size but on average a gene consists of about a **thousand nucleotides** with their bases arranged in a particular sequence.

Each gene acts as a functional unit by coding (via mRNA) for one polypeptide (or protein); the order of the amino acids in the polypeptide (or protein) is determined by the sequence of the bases in the gene's DNA.

KEY QUESTIONS

1 a) What is a gene?
   b) Where are genes found?
   c) Of what type of chemical molecule is a chromosome mainly composed?

2 a) On average, how many DNA nucleotides make up a gene?
   b) Briefly state how a gene exerts its effect.

CONTROL OF GENE ACTION

Some proteins are only required by a cell under certain circumstances. To prevent resources being wasted, the genes that code for these proteins are **'switched on'** and **'switched off'** as required.

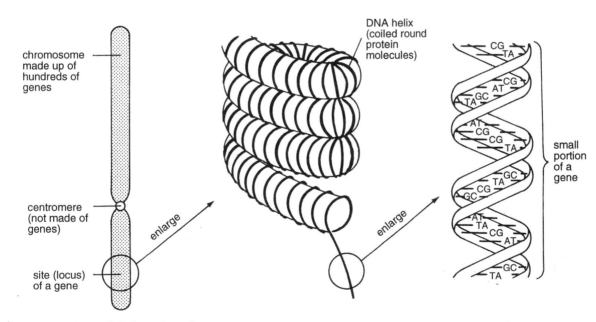

Figure 7.1: A simplified version of gene structure

chromosome made up of hundreds of genes

centromere (not made of genes)

site (locus) of a gene

enlarge

DNA helix (coiled round protein molecules)

enlarge

small portion of a gene

CG
TA
CG
AT
GC
TA
AT
CG
CG
TA
GC
TA
CG
GC
AT
TA
CG
AT
GC
TA

OPERON THEORY OF GENE ACTION

OBSERVATIONS

When the bacterium *Escherichia coli* is placed in growth medium containing glucose, it undergoes rapid cell division (see figure 7.2).

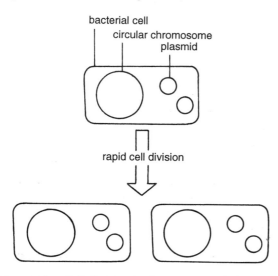

Figure 7.2: Cell division in the presence of glucose

When *E. coli* is placed in growth medium lacking glucose but containing lactose, there is a delay (called the lag phase) before cell division occurs (see figure 7.3).

BACKGROUND INFORMATION

- **Glucose** is a simple monosaccharide sugar (see chapter 1) which is used by *E. coli* for respiration and energy release.
- **Lactose** is a disaccharide sugar composed of a molecule of glucose joined to a molecule of galactose.
- *E. coli* can only make use of the glucose in lactose if it is released from the galactose.
- **β-galactosidase** is an enzyme which catalyses the reaction:

$$\text{lactose} \xrightarrow{\text{β-galactosidase}} \text{glucose} + \text{galactose}$$

- *E. coli* produces β-galactosidase when lactose is present in its food but fails to do so when lactose is absent.

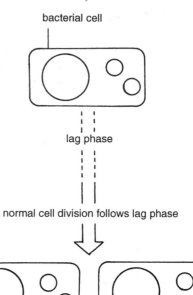

Figure 7.3: Cell division in the presence of lactose

- *E. coli's* chromosome has a gene which codes for β-galactosidase.
- This gene is somehow 'switched on' in the presence of lactose and 'switched off' in the absence of lactose.

OPERON THEORY

It is thought that β-galactosidase is coded for by a **structural** gene which is controlled by a **promoter** gene and an **operator** gene (see figure 7.4). The promoter gene is the site of attachment of the enzyme RNA-polymerase needed for the transcription of the structural gene. These three genes make up the **'lac' operon**. (The promoter and operator genes are known as **control sites** since they are not transcribed but act as sites for the action of molecules.)

Absence of lactose

Environments inhabited by *E. coli* normally contain glucose (not lactose) and so it would be a waste of resources for the bacterium to make the lactose-digestion enzyme when it is not needed. It is prevented from doing so by a **repressor** molecule coded for by a **regulator** gene further along the bacterial chromosome (see figure 7.5).

Figure 7.4: 'lac' operon

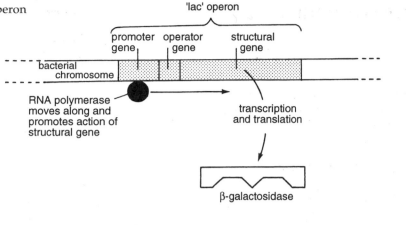

Figure 7.5: Effect of repressor molecule in the absence of lactose

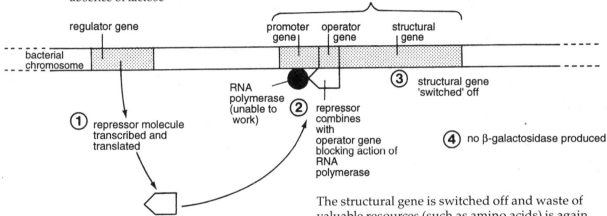

When lactose is absent, the repressor molecule combines with the operator gene preventing the transcription of the structural gene's DNA. The structural gene remains switched off.

Presence of lactose

When the bacterium finds itself in a medium lacking glucose but containing lactose, the events shown in figure 7.6 overleaf take place.

Lactose (the **inducer**) prevents the repressor molecule from combining with the operator gene. The system is no longer blocked and RNA polymerase from the promoter gene is able to go into action and promote the **transcription** of the structural gene. This switching on of the structural gene leads to the production of β-galactosidase.

When all the lactose has been digested by this enzyme, the repressor molecule becomes free again and combines with the operator as before.

The structural gene is switched off and waste of valuable resources (such as amino acids) is again prevented.

The operon theory is supported by much experimental evidence and provides a convincing explanation of gene action.

KEY QUESTIONS

1 Why are some genes found to be switched on at certain times and switched off at other times?

2 **a)** Give the word equation of the biochemical reaction which is controlled by the enzyme β-galactosidase.
b) Under what circumstances does *Escherichia coli* (i) produce β-galactosidase (ii) fail to produce this enzyme?
c) What effect does the presence of lactose in the cell have on the gene which codes for β-galactosidase?

3 Study figure 7.5 overleaf and then answer the following questions.

a) State the THREE types of gene which make up the 'lac operon' in a cell of *E. coli*.

b) Which of these genes (i) is the site of attachment of the enzyme RNA polymerase? (ii) contains the DNA code for the enzyme β-galactosidase? (iii) is affected by the repressor molecule made by the regulator gene?

c) Where is the regulator gene situated in relation to the operon?

d) In the absence of lactose, describe the effect of the repressor molecule on the system.

4 Study figure 7.6 and then answer the following questions.

a) To what molecule does lactose, when present in the cell, become combined without becoming digested?

b) What effect does this have on (i) the operator gene? (ii) the promoter gene's RNA polymerase? (iii) the structural gene?

c) What function does the enzyme β-galactosidase now perform?

d) Describe the series of events that leads to the operon becoming switched off again.

ROLE OF DNA IN CONTROL OF CELL METABOLISM

All the chemical processes that occur in a living organism and keep it alive are known collectively as its **metabolism.** Cell metabolism refers to the biochemical reactions that occur within a cell. Some of these involve the breakdown of a substance (e.g. aerobic respiration – see page 38); others bring about the building up of a substance (e.g. photosynthesis).

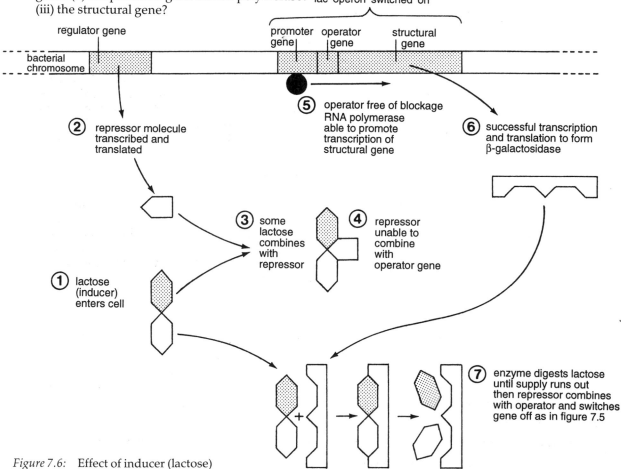

Figure 7.6: Effect of inducer (lactose)

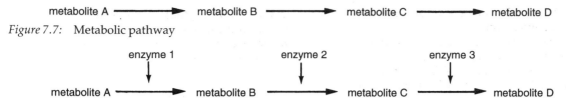

Figure 7.7: Metabolic pathway

Figure 7.8: Enzyme control of metabolic pathway

A **metabolic pathway** normally consists of several stages each of which involves the conversion of one metabolite to another during a breaking-down or building-up process. An imaginary example is shown in figure 7.7.

An **enzyme** is a protein which speeds up the rate of a biochemical reaction. Each stage in a metabolic pathway is controlled by an enzyme (see figure 7.8).

Since each enzyme is made of **protein**, its synthesis is determined by the genetic information held in the DNA of a particular **gene**. Figure 7.9 shows how the cell's metabolism is ultimately under the control of DNA. The genes act by determining the structure of enzymes.

KEY QUESTIONS

1 What is meant by the term *cell metabolism*?

2 a) Copy the following sentence and complete the blanks.
A metabolic pathway normally consists of several _____ each of which involves the conversion of one _____ to another _____ during a breaking-down or _____ process.
b) What type of substance controls the rate at which each stage in a metabolic pathway proceeds?

3 a) Of what substance is every enzyme composed?
b) Of what sub-units is every protein composed?
c) Where in a cell are the instructions for determining the specific sequence of amino acids in a protein kept?
d) By what means are these instructions transferred from DNA to polypeptide (or protein)?
e) How many types of polypeptide (or protein) are determined by one gene?

MUTATION

A **mutation** is a sudden change in the structure or amount of an organism's genetic material. It varies from a tiny change in the DNA structure of a gene (see below) to a large scale alteration in chromosome structure or number (see page 87).

GENE MUTATION

This type of mutation involves a change in one or more of the **nucleotides** in a strand of DNA as shown in figure 7.10. (An insertion involving an extra nucleotide whose base type is the same as the one next to it on the mutated DNA strand is also called a duplication).

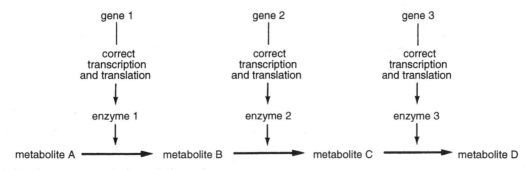

Figure 7.9: Genetic control of metabolic pathway

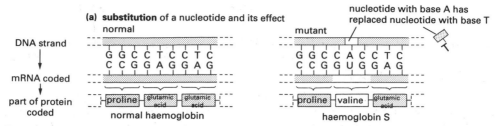

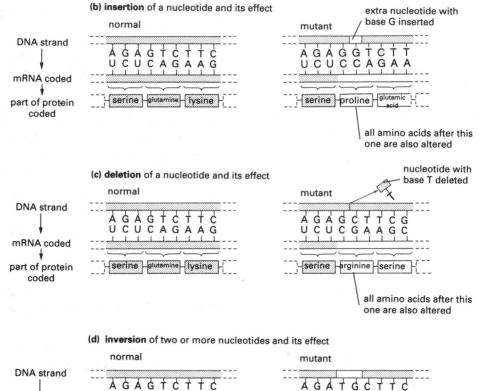

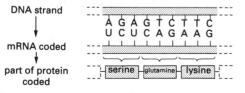

Figure 7.10: Types of gene mutation

In each of the examples in figure 7.10, one or more codons for one or more particular amino acids have become altered leading to a change in the protein that is synthesised.

For a protein to work properly it must have the correct sequence of amino acids. **Substitution** and **inversion** only bring about a **minor** change (i.e. one different amino acid) and often the organism is affected only slightly or not at all.

Insertion and **deletion**, on the other hand, lead to a **major** change since each causes a large portion of the gene's DNA to be misread. The protein

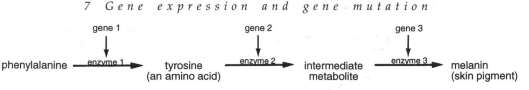

Figure 7.11: Normal fate of phenylalanine

produced differs from the normal protein by many amino acids and it is usually non-functional.

If such a protein is an enzyme which catalyses an essential step in a metabolic pathway, then the pathway becomes disrupted. An intermediate metabolite may accumulate with disastrous results, as follows.

Phenylketonuria

Phenylalanine and tyrosine are two amino acids that humans obtain from protein in their diet. Under normal circumstances, excess phenylalanine follows the pathway shown in figure 7.11.

Phenylketonuria (**PKU**) is a hereditary disorder caused by a genetic defect which disrupts this pathway. An affected person inherits the **mutated form** of gene 1 which is unable to code for enzyme 1 in the pathway. Phenylalanine is therefore no longer converted to tyrosine. Instead it becomes changed into poisons which affect brain cells and reduce mental development.

One of these chemicals is excreted in urine allowing the disorder to be diagnosed. Thanks to widespread screening of newborn babies, followed by a lifelong phenylalanine-restricted diet for sufferers, its worst effects have been reduced to a minimum.

Albinism

Owing to a different mutation, **albinos** are unable to make enzyme 3 in the pathway shown in figure 7.11 and therefore fail to produce **melanin.** Complete lack of this pigment causes an albino to have pink skin (which fails to tan) and white hair. Albinos are perfectly normal in every other way and suffer no medical problems provided that they avoid exposure to ultraviolet radiation.

KEY QUESTIONS

1 What is meant by the term *mutation?*

2 **a)** With reference to nucleotides, describe how each of the following gene mutations comes about: (i) substitution; (ii) insertion; (iii) duplication; (iv) deletion; (v) inversion.
b Which of these gene mutations lead to (i) a change in one amino acid only in the protein chain? (ii) a change in many amino acids in the protein chain?

3 **a)** Name the TWO gene mutations in figure 7.10 that are more likely to lead to the production of non-functional proteins.
b) If an enzyme which catalyses an essential step in a pathway is rendered non-functional following a mutation, what happens to the pathway?

4 **a)** What happens to an excess of the amino acid phenylalanine in a normal human cell?
b) Why is a sufferer of phenylketonuria (PKU) unable to deal with an excess of phenylalanine?
c) What happens to this phenylalanine in the cells of a PKU sufferer?
d) What measures are taken nowadays to reduce the worst effects of PKU to a minimum?

5 **a)** Why are albinos unable to make enzyme 3 in the pathway shown in figure 7.11?
b) What effect does this have on their skin and hair colour?
c) What measures must be taken by albinos to avoid potential problems caused by lack of melanin?

EXERCISES

1 Match the terms in list **X** with their descriptions in list **Y.**

list X	list Y
1) control sites	**a)** stretch of DNA which codes for repressor molecule which may bind to operator gene
2) inducer	**b)** molecule coded for by regulator gene which can bind with operator gene
3) operator gene	**c)** stretch of DNA which codes for a functional protein (e.g. an enzyme)

4) promoter gene

d) molecule which prevents repressor molecule from blocking the operon

5) regulator gene

e) stretch of DNA which acts as a site for action of repressor molecule

6) repressor

f) stretches of DNA such as promoter and operator genes which are not transcribed

7) structural gene

g) stretch of DNA which acts as a site of attachment of RNA-polymerase

2 The right hand part of figure 7.1 on page 65 shows the base sequence of a small portion of a gene. If the complete gene is 1000 bases long, what proportion of it is shown in the diagram?

3 Put the following into order of size beginning with the smallest: chromosome, nucleotide base, gene.

4 Decide whether each of the following statements is true or false and then use T or F to indicate your choice. Where a statement is false, give the word(s) that should have been used in place of the word(s) in **bold print**.
a) A unit of heredity which controls an inherited characteristic is called a **chromosome**.
b) A gene consists of a section of nucleic acid called **RNA**.
c) A chromosome is composed of nucleic acid arranged as a **double helix**.
d) A gene contains a long chain of molecules called **bases** whose sequence determines the genetic information present.

5 Figure 7.12 shows a bacterium's 'lac' operon switched on.
a) Which gene coded for the repressor molecule?
b) (i) Which gene does the repressor molecule normally combine with?

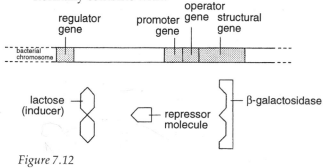

Figure 7.12

(ii) Why is the repressor unable to play this role in the situation shown in the diagram?
c) Which TWO genes are called the control sites?
d) Which gene expresses itself as β-galactosidase?
e) Predict what will happen to the system when all of the lactose has been digested to simple sugars by β-galactosidase.

6 Match the terms in list **X** with their descriptions in list **Y**.

list X	list Y
1) albinism	**a)** gene mutation involving the exchange of one nucleotide for another in the DNA chain
2) deletion	**b)** gene mutation involving the addition of an extra nucleotide to the DNA chain
3) enzyme	**c)** substance that takes part in a metabolic process by being the product of one enzyme and the substrate of another
4) insertion	**d)** the sum of all the chemical reactions occurring within a living organism
5) inversion	**e)** gene mutation involving the loss of one nucleotide from the DNA chain
6) metabolism	**f)** a sudden change in an organism's genetic material
7) metabolite	**g)** a genetically inherited condition in which the sufferer lacks the enzyme needed to break down phenylalanine
8) mutation	**h)** protein molecule which acts as a biological catalyst by speeding up a biochemical reaction
9) phenylke-tonuria	**i)** gene mutation involving the reversal of the order of two nucleotides in the DNA chain
10) substit-ution	**j)** a genetically inherited condition in which the sufferer lacks the enzyme needed to make melanin

7 What name is given to the type of gene mutation illustrated in figure 7.13?

8 What name is given to the type of gene mutation where one incorrect nucleotide occurs in place of the correct nucleotide in a DNA chain?

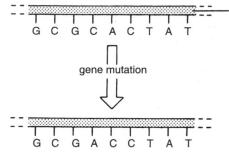

Figure 7.13

9 What name is given to the type of gene mutation illustrated in figure 7.14?

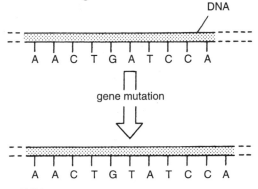

Figure 7.14

10 What name is given to the type of gene mutation where a nucleotide is permanently lost from the DNA chain and not replaced by another?

11 Which of the following gene mutations BOTH lead to a major change which causes a large portion of the gene's DNA to be misread?
A substitution and deletion
B deletion and insertion
C insertion and inversion
D inversion and substitution
(Choose ONE correct answer only.)

12 In the following four telegrams, a small error alters the sense of the message. To which type of gene mutation is each of these equivalent?
a) Intended: She ordered boiled rice.
   Actual: She ordered boiled ice.
b) Intended: He walked to the pillar box.
   Actual: He talked to the pillar box.
c) Intended: She untied the two ropes.
   Actual: She united the two ropes.
d) Intended: He put a quid in his pocket.
   Actual: He put a squid in his pocket.

13 In figure 7.15, a gene mutation has occurred in the abnormal situation.

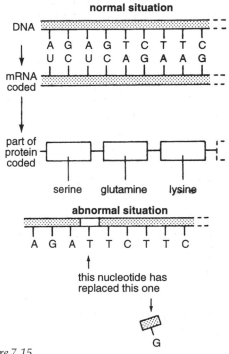

Figure 7.15

a) Name this type of gene mutation.
b) Draw the abnormal DNA as shown and then continue the diagram through mRNA and on to protein to show how this mutation would affect the part of the protein molecule being synthesised.

14 Figure 7.16 (overleaf) shows part of a metabolic pathway that occurs in humans. Each stage is controlled by an enzyme. Some of the stages have been given a letter.
a) Explain how a gene mutation can lead to a blockage in such a pathway.
b) Identify the letter that represents the point of blockage that leads to each of the following disorders:
(i) phenylketonuria
(ii) albinism
(iii) alcaptonuria (characterised by an accumulation of homogentisic acid which is excreted in urine and turns black in light).

15 The graph in figure 7.17 (overleaf) shows the effects of a phenylalanine meal on a normal person and on a sufferer of phenylketonuria (PKU).

a) Explain the initial rise in level of tyrosine in the normal person.

b) Why does the PKU sufferer not show a similar increase?

c) Why does the level of tyrosine in the normal person fall after two hours?

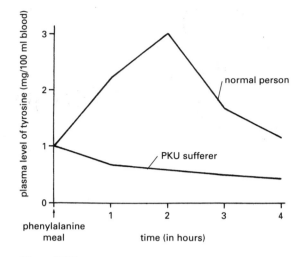

Figure 7.17

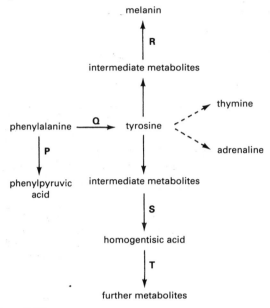

Figure 7.16

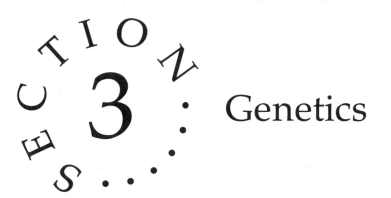

3 : Genetics

8 GENETIC IMPORTANCE OF MEIOSIS

HAPLOID AND DIPLOID CELLS AND CHROMOSOME NUMBER

Every species of plant and animal has a **characteristic number** of **chromosomes** (the chromosome complement) present in the nucleus of each of its cells.

A **diploid** cell (e.g. a zygote as shown in figure 8.1) has a double set of chromosomes (two of each type) which form pairs. A **haploid** cell (e.g. a sex cell as shown in figure 8.1) has a single set of chromosomes (one of each type).

The chromosome complement of a haploid cell is represented by the letter n and that of a diploid cell by 2n. Some examples are shown in table 8.1

Since the members of each pair of chromosomes present in a diploid cell match one another gene

organism	number of chromosomes present in haploid cell (n)	number of chromosomes present in diploid cell (2n)
fruit fly	4	8
buttercup	7	14
onion	8	16
rice	12	24
clover	16	32
cat	19	38
human	23	46
horse	33	66

Table 8.1   Chromosome complements

for gene, they are said to be **homologous.** They may however possess different **alleles** (forms) of the same gene.

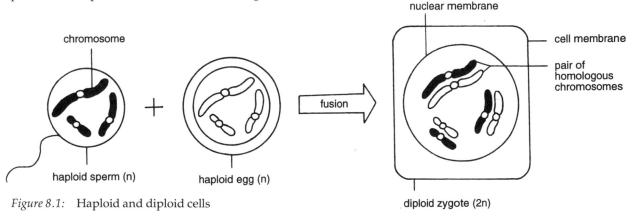

Figure 8.1:   Haploid and diploid cells

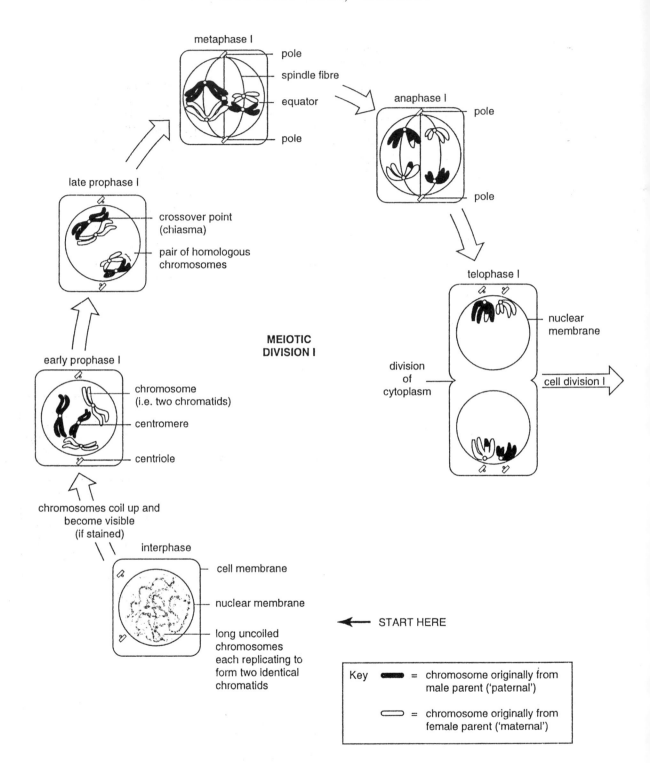

Figure 8.2: Process of meisosis

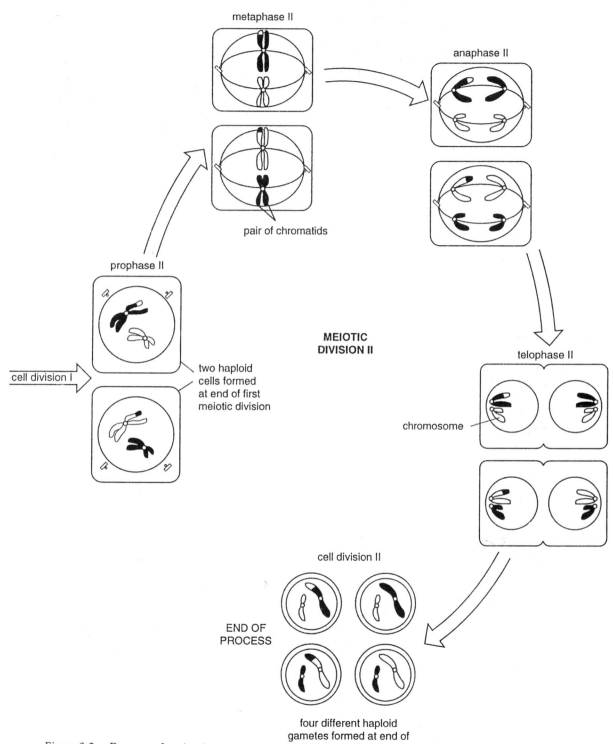

metaphase II

anaphase II

pair of chromatids

prophase II

cell division I

two haploid
cells formed
at end of first
meiotic division

MEIOTIC
DIVISION II

telophase II

chromosome

cell division II

END OF
PROCESS

four different haploid
gametes formed at end of
second meiotic division

Figure 8.2:   Process of meisosis

1 What term is used to refer to the characteristic number of chromosomes possessed by a species of plant or animal?

2 Where in a cell are the chromosomes located?

3 **a)** What term is used to describe a cell such as a gamete which has a single set of chromosomes?
b) What symbol is used to represent this single chromosome complement?

4 **a)** What term is used to describe a cell such as a zygote which contains a double set of chromosomes?
b) What symbol is used to represent this double chromosome complement?

5 How many chromosomes would be present in a
a) human sperm?
b) fruit fly egg?
c) clover pollen grain?

6 State ONE feature that the members of a pair of homologous chromosomes have in common.

PROCESS OF MEIOSIS

Meiosis is a form of nuclear division which results in the production of four haploid (n) gametes from one diploid (2n) gamete mother cell. It occurs at specific sites in living organisms (see table 8.2).

site of meiosis	diploid gamete mother cell	haploid gametes formed
testis of animal	sperm mother cell	sperm
ovary of animal	egg mother cell	eggs (ova)
anther of flowering plant	pollen mother cell	pollen
ovary of flowering plant	egg mother cell	eggs (ovules)

Table 8.2   Sites of meiosis

Meiosis involves two consecutive cell divisions. The gamete mother cell divides into two cells and these then divide again. Figure 8.2 refers to a gamete mother cell containing four chromosomes (two of 'paternal' and two of 'maternal' origin).

During **interphase**, each chromosome replicates forming two identical chromatids. Therefore when the nuclear material becomes visible (on staining)

during **early prophase I**, each chromosome is seen to consist of two chromatids attached at a centromere.

Homologous chromosomes **pair up** during prophase I and come to lie alongside one another so that their centromeres and genes match exactly. While in this paired state, the chromosomes become even shorter and thicker by coiling up.

Next the members of each homologous pair begin to repel one another and move apart except at points called **chiasmata** (singular = chiasma) where they remain joined together. It is here that exchange of genetic material occurs by two chromatids twisting around one another and 'swapping' portions. This exchange is called **crossing over** (see also page 80).

At **metaphase I**, the nuclear membrane disappears, spindle fibres form and homologous pairs become arranged on the equator of the cell. The arrangement of each pair relative to any other is **random** (see also page 79).

On contraction of the spindle fibres during **anaphases I**, one chromosome of each pair moves to one pole and its homologous partner moves to the opposite pole.

During **telophase I**, a nuclear membrane forms round each group of chromosomes. This is followed by division of the cytoplasm resulting in the formation of two haploid cells.

Each of these haploid cells now undergoes the second meiotic division as shown in figure 8.2. Single chromosomes (each made of two chromatids) line up at each equator at **metaphase II**. On separation from its partner during **anaphase II**, each chromatid is regarded as a chromosome.

Each of the four gametes formed contains **half** the number of chromosomes present in the original gamete mother cell.

1 Briefly state what is meant by the term *meiosis*.

2 Name ONE site in an animal and ONE site in a plant where meiosis occurs.

3 How many cell divisions does the process of meiosis involve?

4 Arrange the following stages of meiosis into the correct order starting with interphase: anaphase I, anaphase II, interphase, metaphase I,

arrangement 1

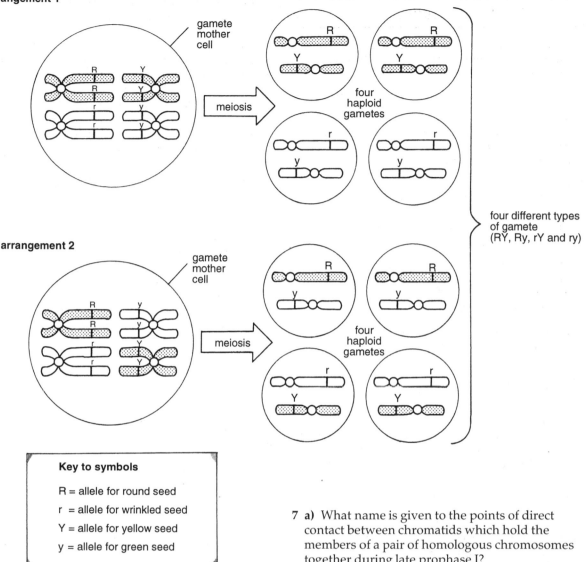

Key to symbols

R = allele for round seed

r = allele for wrinkled seed

Y = allele for yellow seed

y = allele for green seed

Figure 8.3: Independent assortment of chromosomes

metaphase II, prophase I, prophase II, telophase I, telophase II.

5 In what way do chromosomes change during interphase?

6 a) During which stage of meiosis do homologous chromosomes come to lie alongside one another and form pairs?
b) In what way does the appearance of each chromosome become altered at this stage?

7 a) What name is given to the points of direct contact between chromatids which hold the members of a pair of homologous chromosomes together during late prophase I?
b) What may happen at these points of contact?

8 a) Where do homologous pairs of chromosomes become arranged during metaphase I?
b) Rewrite the following sentence and complete the blank.
At metaphase I the arrangement of each homologous pair relative to any other homologous pair is _____ .

9 a) Which structures contract during anaphase I?
b) What effect does such contraction have on the members of each homologous pair of chromosomes?

10 a) Is a gamete mother cell haploid or diploid?
b) Are the products of the first meiotic division haploid or diploid?
c) Are the products of the second meiotic division haploid or diploid?

11 Give ONE difference between metaphase I and metaphase II.

12 Compare the chromosomal number of the original gamete mother cell with that of each of the four gametes formed by meiosis.

RANDOM (INDEPENDENT) ASSORTMENT OF CHROMOSOMES

When homologous pairs of chromosomes line up at the equator during the first meiotic division, the final position of any one pair is **random** relative to any other pair.

Figure 8.3 shows a gamete mother cell in a pea plant (where only two homologous pairs have been drawn). There are two ways in which the pairs can become arranged. Subsequent meiotic divisions bring about **random (independent) assortment** of chromosomes. This gives rise to 2^2 (i.e. 4) different genetic combinations in the gametes.

The larger the number of chromosomes present, the greater the number of possible combinations. For example, a human egg mother cell with 23 homologous pairs has the potential to produce 2^{23} (i.e. 8388608) different combinations.

LINKAGE AND CROSSING OVER

Figure 8.4 shows a homologous pair of chromosomes from a gamete mother cell of a fruit fly. For the sake of simplicity the chromosomes have been drawn as straight 'rods'. In reality, chromosomes are soft flexible structures.

In the example shown in figure 8.4, the genes for wing length and body colour are on the same chromosome and are therefore said to be **linked**. The genes on a chromosome are said to make up a **linkage group**.

Only the genes for wing type and body colour are considered in figure 8.4. The 'paternal' chromosome has the long wing and grey body alleles; the 'maternal' chromosome has the tiny (vestigial) wing and black body alleles for these two genes.

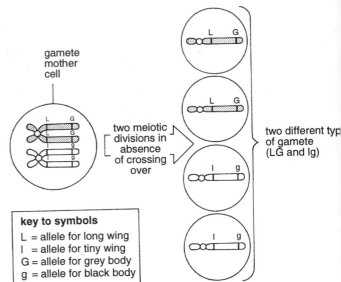

key to symbols
L = allele for long wing
l = allele for tiny wing
G = allele for grey body
g = allele for black body

Figure 8.4: No crossing over

In the absence of crossing over, the long wing and grey body alleles will be transmitted together and pass into 50% of the gametes at meiosis; the tiny wing and black body alleles will be transmitted together and pass into the other 50% of the gametes.

However the members of such a linkage group can be separated if **crossing over** occurs between adjacent chromatids at meiosis (see figure 8.5). In this example, four different types of gamete would be produced and variation would therefore be further increased.

CROSS-OVER FREQUENCY

Figure 8.6 shows a homologous pair of chromosomes bearing three genes A/a, B/b and C/c. In this case allele A is linked with B and C; and allele a is linked with b and c. Since crossing over occurs at random at any point along a chromosome, crossing over tends to occur less frequently between two genes that are close together (e.g. A/a and B/b) than between two genes that are further apart (e.g. B/b and C/c).

The broken lines in figure 8.7 represent five **cross-overs** occurring at random along this homologous pair of chromosomes. Only cross-over 1 would break the linkage between genes A/a

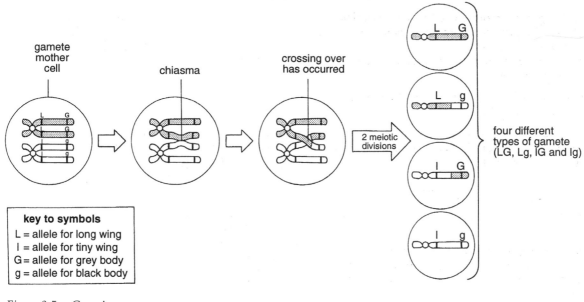

key to symbols
L = allele for long wing
l = allele for tiny wing
G = allele for grey body
g = allele for black body

Figure 8.5: Crossing over

and B/b; however any one of cross-overs 2–5 would break the linkage between genes B/b and C/c. Thus the further apart the two genes, the higher the cross-over frequency between them.

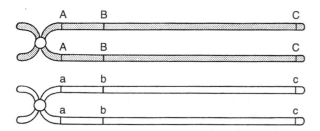

Figure 8.6: Three linked genes

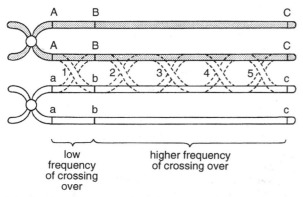

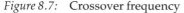

low frequency of crossing over

higher frequency of crossing over

Figure 8.7: Crossover frequency

VARIATION

Much of the variation that exists amongst the members of a species is inherited and determined by alleles of genes.

MEIOSIS

During meiosis, **new combinations** of existing alleles arise by random assortment and crossing-over. These processes therefore increase **variation** within a species.

In addition, meiosis is the means by which complete haploid sets of chromosomes are deposited into gametes. These sex cells then bring together two different haploid versions of a species' genetic blueprint at fertilisation. This mixing of part of one parent's genotype with that of another during sexual reproduction produces a new individual which is **genetically different** from both its parents and from all other members of the species.

Such variation is of great importance because it helps the species to adapt to a changing environment. Imagine, for example, that a new disease appears. If great genetic variation exists amongst the members of the species, then there is a good chance that some of them will be resistant and survive. If they were all identical and susceptible to the disease, the whole species would be wiped out.

KEY QUESTIONS

1 Rewrite the following sentence and complete the blanks using the words in brackets. When _____ pairs of chromosomes line up at the _____ during _____ I of meiosis, the final position of any one pair relative to any other pair is _____ . (equator, random, homologous, metaphase)

2 Imagine a gamete mother cell which contains two homologous pairs.
a) In how many different ways can these pairs become arranged relative to one another at metaphase I?
b) In the absence of crossing-over, how many different genetic combinations of these chromosomes can arise in the gametes following the second meiotic division?
c) Copy the following sentence and complete the blanks. This production of various genetic combinations occurs as a result of _____ _____ of chromosomes.
d) Predict what will happen to the number of different genetic combinations produced as the number of chromosomes increases.

3 What term is used to refer to two genes present on the same chromosome which are transmitted together into gametes at meiosis?

4 **a)** What is a linkage group?
b) By what means can an allele of one gene be separated from an allele of another gene in the same linkage group?
c) What effect does this breaking up of a linkage group normally have on the number of genetically different gamete types formed during meiosis?

5 Copy the following sentence and complete the blanks.
Random _____ and _____ are two ways in which meiosis provides the opportunity for new _____ of the existing inherited _____ to arise.

6 Why is it important that the members of a species are not all identical?

EXERCISES

1 Match the terms in list **X** with their descriptions in list **Y**.

list X

1) centromere
2) chiasma
3) chromatid
4) chromosome
5) chromosome complement
6) diploid
7) gene
8) haploid
9) homologous pair
10) meiosis

list Y

a) one of the two longitudinal sub-units of a duplicated chromosome
b) possessing two sets of chromosomes
c) the characteristic number of chromosomes typical of a species
d) cross-shaped arrangement of two chromatids at a point of crossing-over
e) small region of chromosome which becomes associated with spindle fibres during meiosis
f) unit of heredity occupying a specific site on a chromosome
g) a form of nuclear division producing four haploid gametes from a diploid cell
h) two chromosomes identical in size and matching one another gene for gene (although alleles may differ)
i) possessing one set of chromosomes
j) thread-like structure composed of genes and found in the nucleus of a cell

2 Cut up a copy of figure 8.2 and practise arranging the stages of meiosis into the correct order.

3 With reference to the cell shown in figure 8.8, state the number of:
a) chromatids present;
b) centromeres present;
c) chromosomes present;
d) pairs of homologous chromosomes present;
e) chromosomes that would be present in each gamete produced.

4 Rewrite the following sentences and complete the blanks.
Compared with a haploid cell, a diploid cell contains _____ as many chromosomes. The chromosome complement of a haploid cell

egg mother cell
at start of meiosis

Figure 8.8

is often represented as _____ and that of a
_____ cell as 2n.

Exercises 5–10 are multiple choice items. In each case you should choose ONE correct answer only.

5 Figure 8.9 shows the formation of an animal zygote.

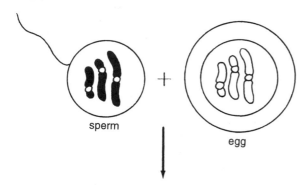

sperm

egg

Figure 8.9   zygote

This zygote contains
A 6 chromatids and is haploid.
B 6 chromatids and is diploid.
C 3 pairs of chromosomes and is haploid.
D 3 pairs of chromosomes and is diploid.

6 In which of the following do BOTH structures contain cells which divide by meiosis?
A anther and stigma
B stamen and style
C ovary and anther
D stigma and stamen

7 Each pollen mother cell in a white clover plant contains 32 chromosomes. The number of chromosomes present in a shoot tip cell would be
A 8.   B 16.   C 32.   D 64.

8 Which of the diagrams in figure 8.10 correctly shows a pair of homologous chromosomes at the start of meiosis?

A   B

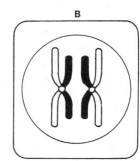

C   D

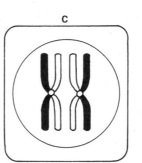

Figure 8.10

9 Figure 8.11 shows a cell undergoing meiosis. Which of the diagrams in figure 8.12 shows the next stage in the process?

chiasma —

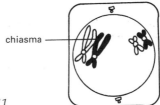

Figure 8.11

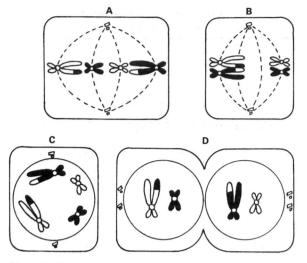

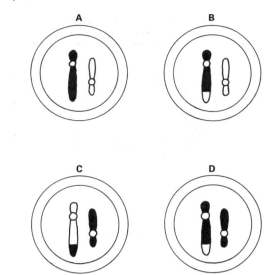

Figure 8.12

Figure 8.14

10 Figure 8.13 shows a cell undergoing meiosis. Assume that crossing over occurs only at the chiasma indicated.

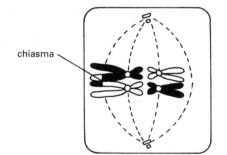

Figure 8.13

Which of the gametes shown in figure 8.14 will NOT be formed from this cell?

11 Four steps that occur during meiosis are given in the following list:
a) complete separation of chromatids
b) pairing of homologous chromosomes
c) lining up of paired chromosomes on equator
d) crossing-over between chromatids
Arrange these steps in the correct order.

12 Table 8.3 takes the form of a grid containing statements which describe the behaviour of chromosomes during meiosis.

1 homologous pairs become arranged on the equator of the cell	2 each chromosome duplicates forming two identical chromatids
3 homologous chromosomes are pulled to opposite poles of the cell	4 each chromatid becomes separated from its partner
5 homologous chromosomes pair up and lie alongside one another	6 single chromosomes (each made of two chromatids) line up at each equator

Table 8.3

Which description applies to:
a) interphase?
b) prophase I?
c) metaphase I?
d) anaphase I?
e) metaphase II?
f) anaphase II?

13 Match diagrams 1–4 in figure 8.15 with diagrams a–d.

14 Match the terms in list X with their descriptions in list Y

list X	**list Y**
1) alleles	**a)** association and transmission together of a group of genes from generation to generation

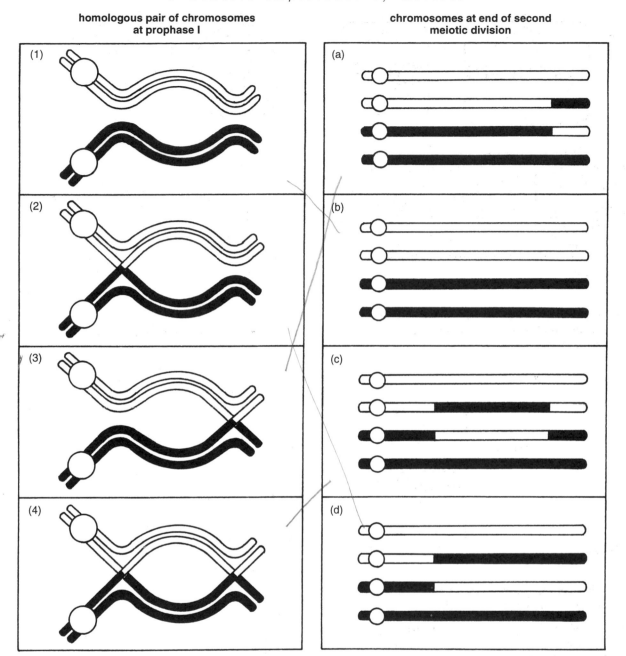

**homologous pair of chromosomes
at prophase I**

**chromosomes at end of second
meiotic division**

Figure 8.15

2 crossing-over

b) arrangements of homologous pairs of chromosomes which allow their independent segregation at meiosis

3 inherited variation

4) linkage

c) group of genes located together on the same chromosome

d) breaking and rejoining of adjacent chromatids leading to exchange of genetic material

5) linkage group

e) differences that exist amongst the members of a species which are determined by genes

6) random assortment of chromosomes

f) alternative forms of a gene that occur at the same position on a pair of homologous chromosomes

15 Sandra and Tracy are twin sisters yet they differ in several ways including eye colour, blood group and tongue-rolling ability. Each twin has received half of her chromosomes from each of the same two parents yet the twins are non-identical. Explain why in terms of meiosis.

16 Figure 8.16 shows a gamete mother cell of maize at metaphase I of meiosis. Only two homologous pairs of chromosomes have been drawn. One of these bears the gene for grain type (alleles – F for full, f for shrunken); the other the gene for grain colour (alleles – P for purple, p for yellow).
a) Using only the letters given, state the type(s) of gamete that would result from this cell following meiosis (assuming that no crossing-over had occurred).
b) (i) Draw a simple diagram to show an alternative way in which the homologous pairs could become arranged at the equator during meiosis.
(ii) State the type(s) of gamete that would result from this second arrangement following meiosis (assuming that no crossing-over had occurred).

17 Copy and complete table 8.4.

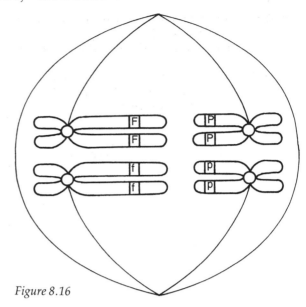

Figure 8.16

18 Figure 8.17 shows a homologous pair of chromosomes during meiosis. The letters refer to three genes.

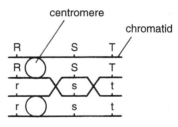

Figure 8.17

Which of the diagrams in figure 8.18 correctly represents the final products of the second meiotic division? (Choose ONE answer only.)

chromosome number		number of different combinations of homologous chromosomes that can arise in gametes following random assortment
diploid	haploid	
4	2	$2^2 = 4$
	3	$2^3 = 8$
8		$2^4 =$
	5	$= 32$
12		2^6
		$2^{23} = 8388608$

Table 8.4

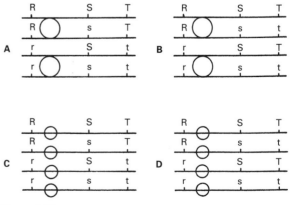

Figure 8.18

19 Table 8.5 refers to 800 eggs made by an animal with the genotype CcDdEe.

a) State the genotype of the fourth type of egg in the table.

b) Are the three genes linked, or are they found on three different chromosomes? Explain your choice of answer.

genotype of egg	number produced
CDE	305
Cde	98
cDE	94
	303

Table 8.5

20 Figure 8.19 shows a pair of homologous chromosomes during meiosis.

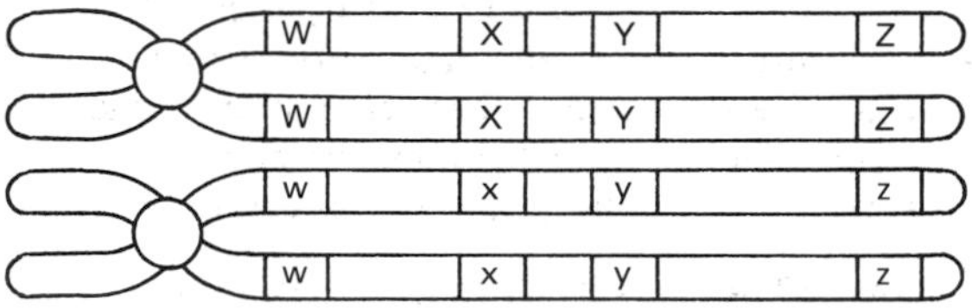

Figure 8.19

Most crossing over will occur between genes
A W and X.
B X and Y.
C W and Z.
D Y and Z.
(Choose ONE correct answer only.)

9 GENETIC CONSEQUENCES OF CHROMOSOMAL CHANGE

CHANGE IN STRUCTURE OF ONE CHROMOSOME

A **chromosome** is made up of many units of heredity called genes. Any change in structure of the chromosome results in the number or sequence of these genes becoming altered.

Such a change is most likely to occur when chromatids break and rejoin during crossing-over at meiosis.

INVERSION

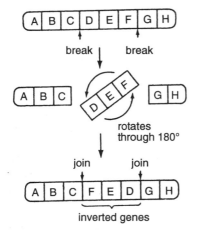

Figure 9.1: Inversion

The chromosome undergoing inversion breaks in two places as shown in figure 9.1. The segment between the two breaks turns round before joining up again. This brings about a **reversal** of the normal sequence of genes in the affected section of chromosome.

When a chromosome which has undergone inversion meets its normal non-mutated homologous partner at meiosis, the two have to form a complicated loop in order to pair up. If crossing-over then occurs within the loop, chromatids find it difficult or even impossible to become separated. Non-viable gametes are often the result.

DUPLICATION

A chromosome undergoes this type of change when a segment of its homologous partner becomes attached to one end of the first chromosome or inserted somewhere along its length as shown in figure 9.2. This results in a set of genes being **repeated.**

In some cases, one of the duplicated genes may change (by mutation) and evolve a new function.

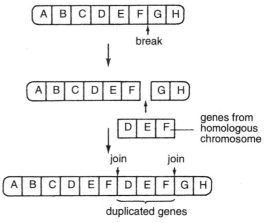

Figure 9.2: Duplication

DELETION

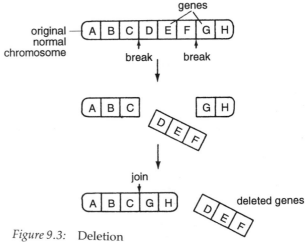

Figure 9.3: Deletion

The chromosome breaks in two places and the segment in between drops out as shown in figure 9.3. The two ends then join up giving a shorter chromosome which **lacks** certain genes.

Deletion normally has a drastic effect on the organism involved. In humans, for example, deletion of part of chromosome 5 leads to the *cri du chat* syndrome. The sufferer is mentally retarded and has a small head with widely spaced eyes. (The condition is so-called because an infant sufferer's crying resembles that of a cat.)

TRANSLOCATION

This involves a section of one chromosome breaking off and becoming attached to another chromosome which is not its partner as shown in figure 9.4.

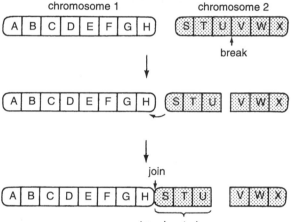

Figure 9.4:   Translocation

This condition usually leads to problems during pairing of homologous chromosomes at meiosis. The gametes formed are often non-viable.

KEY QUESTIONS

1 State FOUR ways in which a chromosome can become changed in such a way that the number or sequence of some of its genes also becomes altered.

2 a) What name is given to a change which involves a chromosome breaking in two places and a segment of genes dropping out?
b) Suggest what would happen to the section of genes that dropped out.

3 a) Name the type of change which involves a chromosome breaking in two places and the affected length of genes be rotated through 180° before becoming reunited with the chromosome.
b) What effect does this change have on the sequence of the genes in the affected segment?

4 a) What name is given to the type of chromosomal change which involves a segment of genes from one chromosome becoming inserted somewhere along the length of its homologous partner?
b) What effect does this change have on the gene pattern of the chromosome which gained the extra genes?

5 a) Name the type of change which involves a section of one chromosome breaking off and joining onto another non-homologous one.
b) What effect does this change have on the number of genes present on each of the affected chromosomes?

CHANGE IN NUMBER OF CHROMOSOMES

Sometimes a change occurs which results in a cell receiving one or more extra chromosomes. Such a change in **number** of chromosomes which varies from one extra chromosome to a full set being added to the normal chromosome complement, occurs as a result of **non-disjunction.**

NON-DISJUNCTION DURING MEIOSIS

When a **spindle fibre** fails during meiosis and the members of one pair of homologous chromosomes fail to become separated, this is called **non-disjunction.** Normal meiosis and non-disjunction are compared in figure 9.5

In this example of non-disjunction (where spindle failure occurs during the first meiotic division), two of the gametes receive an extra copy of the affected chromosome and two gametes lack that chromosome.

On the other hand, if non-disjunction occurs during the second meiotic division and only affects one of the two cells, this results in the production of two normal gametes and two abnormal gametes. One of the abnormal gametes gets an

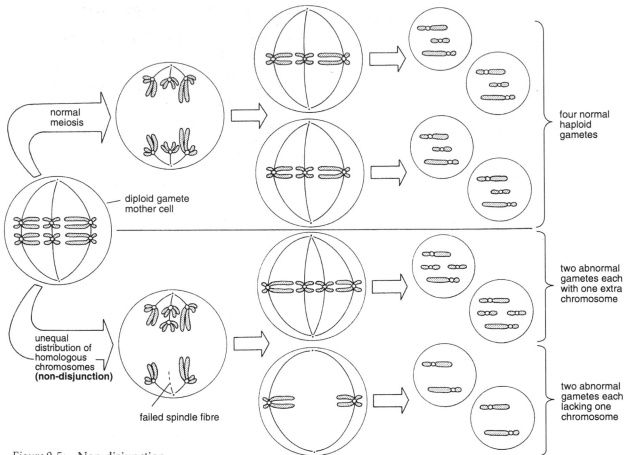

normal meiosis

diploid gamete mother cell

unequal distribution of homologous chromosomes **(non-disjunction)**

failed spindle fibre

four normal haploid gametes

two abnormal gametes each with one extra chromosome

two abnormal gametes each lacking one chromosome

Figure 9.5: Non-disjunction

extra copy of the affected chromosome and the other lacks that chromosome.

Complete non-disjunction during meiosis

When **all** the spindle fibres fail during meiosis and the members of all homologous pairs fail to become separated, this is called **complete non-disjunction.**

In the example shown in figure 9.6, complete non-disjunction occurs during the first meiotic division. This results in the formation of two abnormal diploid gametes.

On the other hand, if total spindle failure occurs at the second meiotic division and affects only one of the two cells, this results in the production of two normal haploid gametes and one abnormal diploid gamete.

Complete non-disjunction during mitosis

Normally a parental cell undergoing mitosis divides to form two daughter cells each with the same chromosome number as itself. However a parental cell can suffer total spindle failure and complete non-disjunction of its chromatids during mitosis. When this happens, it forms only **one daughter cell** which contains **double** the original number of chromosomes.

KEY QUESTIONS

1 What name is given to the process which results in a cell receiving one or more extra chromosomes?

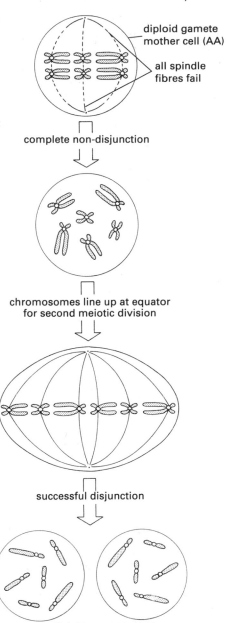

Figure 9.6: Complete non-disjunction

2 Name the type of structure that can fail during meiosis leading to non-disjunction of chromosomes.

3 If one spindle fibre fails at the first meiotic division,

a) how many homologous pairs will be affected?
b) how many gametes will receive an extra chromosome?

4 If one spindle fibre fails at the second meiotic division, how many of the gametes formed will receive an extra chromosome?

5 If all the spindle fibres fail at the first meiotic division,
a) how many homologous pairs will be affected?
b) how many abnormal diploid gametes will be formed?

6 Imagine that during the second meiotic division, one of the cells about to divide suffers complete non-disjunction and the other is unaffected. What types of gamete will be formed and in what numbers?

7 The cell shown in figure 9.7 is undergoing mitosis and is at metaphase.

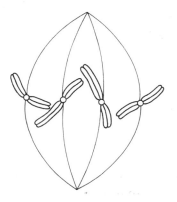

Figure 9.7:

a) Describe the cells that will be formed following normal mitosis.
b) Describe the cell that would be formed following abnormal mitosis involving total spindle failure.

CHROMOSOME NUMBER

DOWN'S SYNDROME

If non-disjunction of **chromosome pair 21** occurs in a human egg mother cell, then one or more

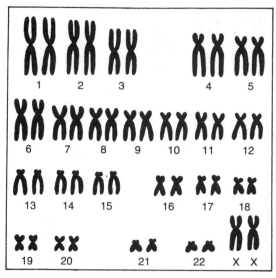

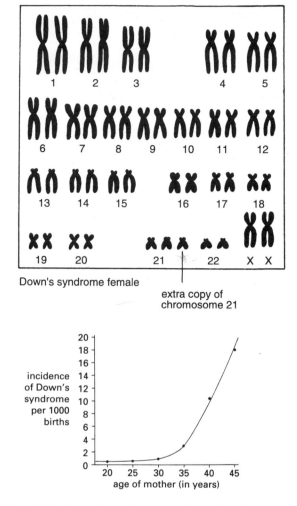

Figure 9.8: Normal and Down' syndrome karyotypes

abnormal eggs (n = 24) may be formed. If one of these is fertilised by a normal sperm (n = 23), this results in the formation of an abnormal zygote (2n = 47). The extra copy of chromosome 21 can be seen in the **karyotype** (display of matched chromosomes) from one of the cells of an affected person (see figure 9.8).

The affected individual suffers from **Down's syndrome** which is characterised by mental retardation and distinctive physical features (such as a short nose and prominent eye lids that slant downwards).

Nearly 80 per cent of the non-disjunctions that lead to Down's syndrome are of maternal origin with frequency being related to maternal age (see figure 9.9). Since egg mother cells of older women seem to be more prone to non-disjunction at meiosis, mothers over the age of thirty-five are advised to have foetal chromosome analysis carried out.

Figure 9.9: Down's syndrome and age of mother

to the formation of eggs possessing only twenty-two chromosomes. What other type of abnormal egg can be formed at the same time?

3 **a)** If a normal human sperm fertilises an egg containing an extra copy of chromosome 21, what is the diploid number of the zygote formed?
b) What name is given to the condition suffered by a person who develops from an abnormal zygote of this type?
c) Name TWO characteristics typical of a sufferer of this condition.

4 What relationship exists between age of mother and incidence of Down's syndrome?

KEY QUESTIONS

1 State the total number of chromosomes present in a normal human egg.

2 Non-disjunction of chromosome pair 21 in a human egg mother cell during meiosis can lead

NON-DISJUNCTION OF SEX CHROMOSOMES

If human **sex chromosomes** are affected by non-disjunction during meiosis, then unusual gametes are formed as shown in figure 9.10.

TURNER'S SYNDROME

If a gamete which possesses no sex chromosomes meets and fuses with a normal X gamete, the zygote formed has the chromosome complement 2n = 45 (44 + XO where O represents the lack of a second sex chromosome). Figure 9.11 shows the karyotype.

An individual with this unusual chromosome complement suffers a condition known as **Turner's syndrome.** Such individuals are always

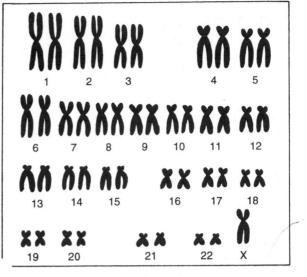

Figure 9.11: Turner's syndrome karyotype

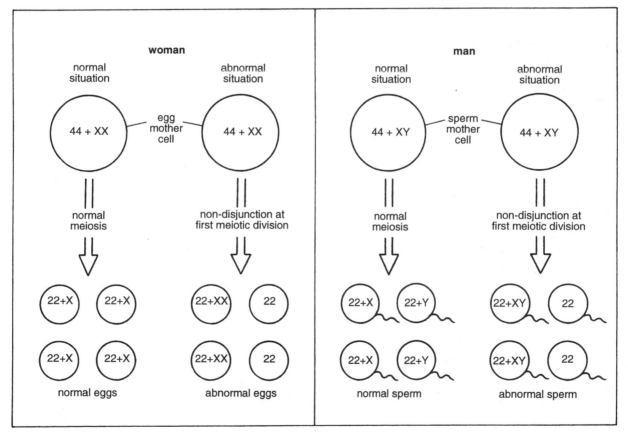

Figure 9.10: Non-disjunction of sex chromosomes

female and short in stature. Since their ovaries do not develop, they are infertile and fail to develop secondary sexual characteristics (e.g. breast development, menstruation etc.) at puberty.

Turner's syndrome occurs with a frequency of about 1 in 2500 female live births.

KLINEFELTER'S SYNDROME

If an XX egg (see figure 9.10) is fertilised by a normal Y sperm or a normal X egg is fertilised by an XY sperm then the zygote formed has the chromosome complement 2n = 47 (44 + XXY). Figure 9.12 shows the karyotype.

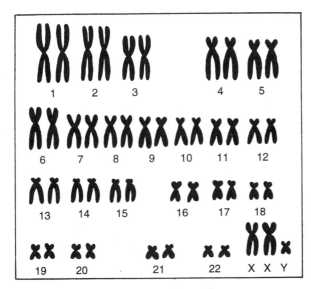

Figure 9.12: Klinefelter's syndrome karyotype

An individual with this unusual chromosome complement suffers a condition known as **Klinefelter's syndrome.** Such individuals are always male and possess male sex organs. However they are infertile since their testes only develop to about half the normal size and fail to produce sperm. The small testes also fail to produce normal levels of testosterone (the male sex hormone) and as a result, male secondary sexual characteristics (e.g. growth of facial hair, deepening of voice etc.) are only weakly expressed. Some sufferers tend to develop small breasts.

Klinefelter's syndrome occurs with a frequency of about 1 in 1000 male live births.

1 State the TWO types of sex chromosome found in human cells.

2 Using letters to indicate sex chromosomes and numbers for the totals of the other chromosomes, give the chromosome complement(s) of:
  a) a normal female adult,
  b) a normal male adult,
  c) a normal egg,
  d) normal sperm (both types).

3 If non-disjunction occurs during the first meiotic division of a human egg mother cell, what types of abnormal egg are formed?

4 If non-disjunction occurs during the first meiotic division of a human sperm mother cell, what types of abnormal sperm are formed?

5 **a)** Give the chromosome complement of the zygote formed when a sperm lacking either sex chromosome fertilises a normal egg.
  b) What name is given to the condition suffered by a person who develops from this abnormal zygote?
  c) State the sex of such a person and give TWO physical features that are typical of the condition suffered.

6 **a)** Give the chromosome complement of the zygote formed when an unusual egg containing two sex chromosomes is fertilised by a normal Y sperm.
  b) What name is given to the condition suffered by a person who develops from this abnormal zygote?
  c) State the sex of such a person and give TWO physical features that are typical of the condition suffered.

POLYPLOIDY

The single haploid set of chromosomes typical of a species is called its **genome.** This is often represented by a symbol (e.g. A). **Polyploidy** is the increase in number of the species' genome by three or more times. It is common against plants but very rare amongst animals.

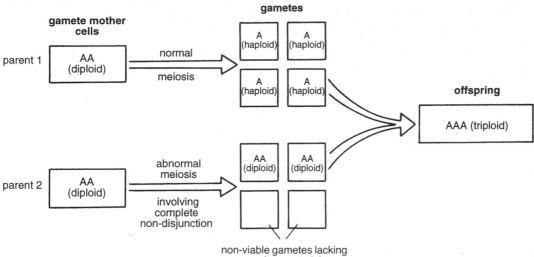

Figure 9.13: Formation of triploid organism

POLYPLOIDY FROM A SINGLE SPECIES

Some polyploids (e.g. AAA, AAAA, etc.) contain more than two genomes all derived from a single species (AA). Such polyploids arise as follows.

All the chromosomes in a gamete mother cell of AA undergo complete non-disjunction (see page 88) during meiosis resulting in the formation of two **diploid gametes**.

If one of these diploid gametes (AA) fuses with a normal haploid gamete (A), a **triploid zygote** (**AAA**) is formed as shown in figure 9.13.

Although capable of mitosis and normal growth,

this triploid organism is **sterile** since its chromosomes cannot form homologous pairs during meiosis. Many varieties of daffodil and other bulb plants are sterile triploids which survive by asexual reproduction.

On the other hand, if a diploid gamete (AA) fuses with another diploid gamete (AA), a **tetraploid zygote** (**AAAA**) is formed as shown in figure 9.14. This organism is capable of normal growth and may also be **fertile** since each of its chromosomes will be able to find a homologous partner at meiosis.

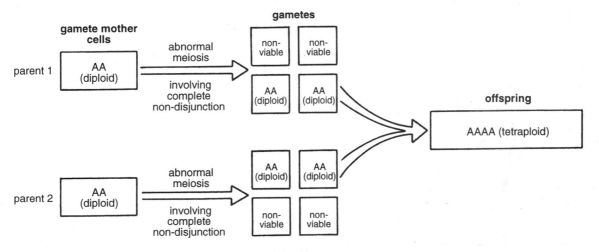

Figure 9.14: Formation of tetraploid organism (AAAA)

POLYPLOIDY FROM MORE THAN ONE SPECIES

Some polyploid plants (e.g. AABB) contain multiple sets of chromosomes derived from more than one species. For example a gamete containing genome A from species AA fuses with a gamete containing genome B from species BB to produce a **sterile hybrid AB.** This organism is sterile because its chromosomes do not match one another and cannot form homologous pairs at meiosis.

However plant AB is capable of mitosis and normal growth and can survive indefinitely by asexual reproduction (e.g. production of runners, bulbs, corms etc.).

Eventually, total spindle failure may occur in one of AB's cells during mitosis leading to the formation of **tetraploid cell, AABB,** as shown in figure 9.15.

Cell AABB divides by normal mitosis and develops into tetraploid plant AABB. On reaching sexual maturity, this type of polyploid plant is **fertile** since all of its chromosomes can find partners and form homologous pairs at meiosis.

Economic significance of polyploidy

Polyploid plants normally have larger cells than their diploid relatives. This results in an **overall increase** in plant size including, most importantly from an economic point of view, seed and fruit size. Many crop plants such as wheat, coffee, apples and tomatoes are polyploid and therefore give bigger yields than their diploid ancestors.

Polyploid plants with an odd number of chromosome sets (e.g. AAA, AAB, ABB etc.) are sterile. However this is useful to humans because the affected plants produce **seedless fruits** e.g. banana.

Where a polyploid is the product of hybridisation, it is often found to show an increase in **vigour** and **resistance to disease** as a result of a combination of characteristics from both of its ancestors.

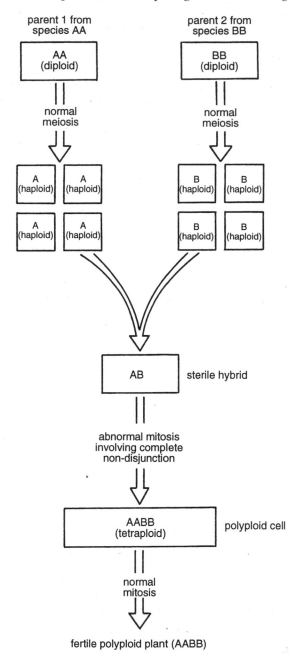

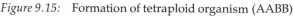

Figure 9.15: Formation of tetraploid organism (AABB)

KEY QUESTIONS

1 What term is used to refer to the single haploid set of chromosomes typical of a species?

2 What is meant by the term *polyploidy?*

3 The chromosome complement of a certain diploid plant species is represented by the symbols BB.
a) State the symbol(s) for: (i) its genome; (ii) a normal haploid pollen grain; (iii) an abnormal diploid egg; (iv) a triploid offspring.
b) What process during meiosis leads to the

production of abnormal diploid eggs?

c) By what means is a triploid offspring produced?

d) Why is a triploid offspring sterile?

e) Briefly describe ONE way in which tetraploid BBBB could be formed.

4 a) State the symbols representing the two separate species from which tetraploid CCDD could have arisen.

b) Using symbols, represent the zygote formed when a pollen grain with genome C fertilises an egg with genome D.

c) Suggest why hybrid CD is found to be sterile.

d) What process must occur for a cell of CD to become CCDD? Does this happen during meiosis or mitosis?

5 Give THREE ways in which polyploid plants can be of economic importance to farmers.

ALTERATION OF PROTEIN STRUCTURE

Each molecule of a **protein** is built up from a large number of sub-units called **amino acids** of which there are about twenty different types. These sub-units are joined together by bonds into long chains (see page 2).

Figure 9.16 shows a molecule of a protein called **haemoglobin** and a close-up of a few of the amino acids found at the end of one of its chains.

A protein is only able to function properly if it contains its own particular combination of amino acids joined together in a specific order. This critical **sequence of amino acids** in a protein is determined by the organism's **genetic code** (see page 58). The genetic code is contained in the organism's chromosomes where the message held by each gene corresponds to the order of amino acids in one protein (or polypeptide) chain.

If a gene becomes altered during a chromosomal change (mutation) then the structure of the protein that it codes for also becomes altered. In most cases such a change results in the protein being unable to carry out its normal function. Only on very rare occasions does the change give the affected organism an advantage.

SICKLE CELL ANAEMIA

Haemoglobin is a protein present in red blood cells. Normal haemoglobin acts as an efficient

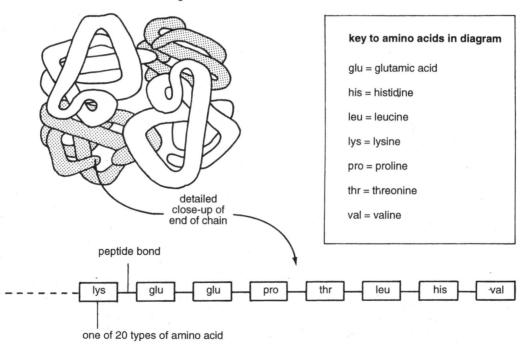

molecule of haemoglobin

detailed close-up of end of chain

peptide bond

lys — glu — glu — pro — thr — leu — his — val

one of 20 types of amino acid

key to amino acids in diagram

glu = glutamic acid

his = histidine

leu = leucine

lys = lysine

pro = proline

thr = threonine

val = valine

Figure 9.16: Structure of haemoglobin

carrier of oxygen to all parts of the human body via the bloodstream. Each molecule of normal haemoglobin consists of many amino acids linked together into a particular sequence determined by the genetic code contained in two genes (see figure 9.17).

A change in one of the genes which code for haemoglobin leads to the formation of an unusual type of haemoglobin (see figure 9.17). This is called **haemoglobin S** and it differs in structure from normal haemoglobin by only **one** amino acid. Although this alteration is tiny it leads to profound changes in the folding and ultimate shape of the haemoglobin S molecule making it a very inefficient carrier of oxygen.

People who have the altered gene on both chromosomes suffer drastic consequences. In addition to all of their haemoglobin being type S which fails to perform the normal function properly, sufferers also possess distorted sickle-shaped red blood cells which stick together and interfere with blood circulation. The result of these problems is severe shortage of oxygen followed by damage to vital organs and, in most cases, early death.

This genetically transmitted condition is called **sickle cell anaemia.** It is very rare in Britain but common in certain parts of Africa.

KEY QUESTIONS

1 a) What name is given to the sub-units of which a molecule of protein is composed?

b) How many different types of these sub-units are found to be present in proteins?

2 a) What aspect of its structure is critical if a protein molecule is to carry out its function properly?
b) What determines this sequence of amino acids in a protein?
c) Where is the genetic code held in a cell?
d) What happens to the structure of a protein if a chromosomal change (mutation) alters the gene that codes for it?

3 a) Which essential protein is not produced in its normal form by a sufferer of sickle cell anaemia?
b) Why is the sufferer unable to make the normal form of the protein?
c) Which protein is made instead?
d) In what way does this protein differ in structure from the normal form?
e) What effect does this change in structure have on the protein's function?

DYSFUNCTION

Dysfunction is the name given to a disturbance or abnormality in function of some part or system of the body. In some cases this occurs as a direct result of an **alteration** in chromosome structure. Sickle cell anaemia, for example, occurs as a consequence of a chromosomal change which affects the structure of haemoglobin, a transport protein.

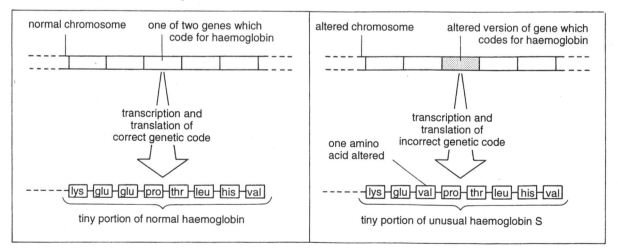

Figure 9.17: Formation of two types of haemoglobin

HAEMOPHILIA

Clotting of blood is the result of a complex series of biochemical reactions involving many essential chemicals. One of these blood-clotting agents is a protein called **factor VIII**. In humans the genetic information for factor VIII is held in a gene carried on the X chromosome.

However this information is rendered useless if the gene is changed by a mutation. A person who inherits the altered genetic material suffers a dysfunctional condition called **haemophilia.** The sufferer's blood takes a very long time (or even fails) to clot resulting in prolonged bleeding from even the tiniest wound. Internal bleeding may occur and continue unchecked leading to serious consequences.

The frequency of haemophilia is 1 in 5000 male births in Britain. It is a sex-linked condition (see page 112).

CYSTIC FIBROSIS

Mucus is the slimy substance secreted by the inner lining of the windpipe and intestine. Mucus is composed of a particular type of protein which makes it thick and slimy and perfectly suited to its roles of protection and lubrication. The genetic information for this protein in mucus is contained in a gene carried on chromosome 7 in humans.

However this information becomes altered by a mutation which affects chromosome 7. People who inherit both copies of chromosome 7 in its altered state are found to make **abnormally thick sticky mucus** which leads to lung congestion and blockage of the pancreatic duct. This dysfunctional condition is called **cystic fibrosis.** It occurs with a frequency of 1 in 2500 in Britain.

Gene therapy

Gene therapy is the **replacement** of defective genes with normal healthy ones. It seems likely that this technique will provide a cure for some types of genetic dysfunction in years to come. For example it may become possible to remove defective cells from the bone marrow of a sufferer of sickle cell anaemia and add a normal haemoglobin gene to each. On being returned to the sufferer's bone marrow these cells would then multiply and produce normal blood.

Similarly it may be possible in the future to use liver cells for gene therapy against haemophilia. Already scientists are working on a method of delivering healthy genes to the lungs of sufferers of cystic fibrosis using an inhaler.

BIOCHEMICAL DYSFUNCTION

Enzymes are biological catalysts which speed up the rate of chemical reactions. Enzymes are made of protein and their structure is determined by genes. Following certain gene mutations, enzymes which control critical steps in biochemical pathways are no longer produced. In each case the pathway becomes disrupted leading to a **biochemical dysfunction** of the body such as phenylketonuria (see page 71).

KEY QUESTIONS

1 In general, what is meant by the term *dysfunction* with respect to the human body?

2 Explain why haemophilia is aptly described as a dysfunctional condition.

3 **a)** What is mucus?
b) What chemical does normal mucus contain that gives it is slimy consistency?
c) Where is the genetic information which controls the formation of this protein held?
d) Under what circumstances is abnormally thick sticky mucus formed?
e) What name is given to the dysfunctional condition caused by this very thick mucus?
f) State TWO problems suffered by a person with this condition.

EXERCISES

1 Match the numbered descriptions in list **X** with the lettered names of chromosomal change in list **Y**.

list X	list Y
1) doubling up of part of a chromosome consisting of several genes	a) non-disjunction
2) transfer of a segment of genes from one chromosome to another non-homologous one	b) deletion

3) increase in chromosome number of cell caused by spindle failure during meiosis or mitosis

c) inversion

4) loss of segment consisting of several genes from a chromosome

d) trans-location

5) reversal of the gene order of a segment of chromosome as a result of two breaks in the same chromosome

e) duplication

2 Figure 9.18 shows a pair of homologous chromosomes during meiosis. What name is given to the type of change that has affected the altered chromosome?

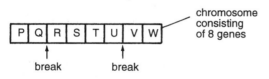
unaltered chromosome
altered chromosome

Figure 9.18

3 If the chromosome shown in figure 9.19 breaks at the points indicated by arrows and the genes between these points become inverted then the resulting order of the genes will be

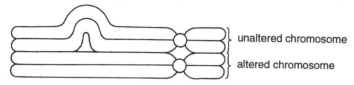
chromosome consisting of 8 genes
break   break

Figure 9.19

A PQUTSRVW
B WVUTSRQP
C PQTURSVW
D VWUTSRPQ
(Choose ONE correct answer only.)

4 Figure 9.20 shows two chromosomes. The lettered regions represent genes.

chromosome 1   chromosome 2

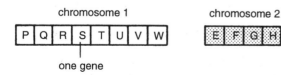
one gene

Figure 9.20

Which of the possible outcomes shown in figure 9.21 would result if a translocation occurred between chromosomes 1 and 2? (Choose ONE correct answer only.)

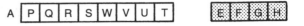
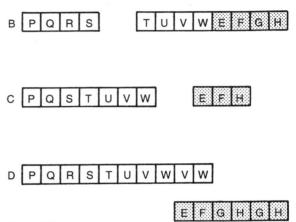

Figure 9.21

5 Figure 9.22 shows two types of chromosome mutation.

These are called

	1	2
A	duplication	deletion
B	duplication	translocation
C	inversion	deletion
D	inversion	translocation

(Choose ONE correct answer only.)

one gene
part of a chromosome
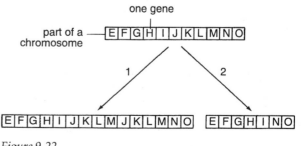

Figure 9.22

100

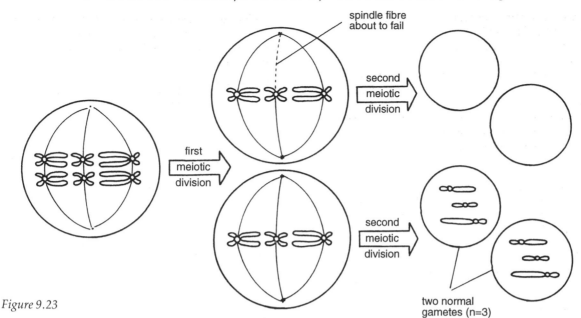

Figure 9.23

two normal
gametes (n=3)

6 Figure 9.23 shows meiosis in a gamete mother cell which has the chromosome complement 2n = 6. In the example shown, one spindle fibre fails during the second meiotic division.

a) Make a simple diagram of the two gametes that would result from the cell that suffers the above spindle failure.
b) Draw a simple diagram of what would have been produced from the same cell if it had suffered complete non-disjunction during the second meiotic division.

7 Match the terms in list **X** with their descriptions in list **Y**.

list X	list Y
1) Down's syndrome	**a)** increase in number of species' genome by three or more times ⁹
2) dysfunction	**b)** condition characterised by chromosome complement 2n = 44 + XO ¹⁰
3) genome	**c)** single complete haploid set of chromosomes typical of a species ₃
4) karyotype	**d)** possesssing four complete sets of chromosomes ₆
5) Klinefelter's syndrome	**e)** condition resulting from non-disjunction of chromosome 21 during meiosis ₁
6) polyploidy	**f)** display of matched chromosomes showing their number, form and size ₄
7) protein	**g)** disturbance or abnormality in function of part of system of body ₂
8) sterile hybrid	**h)** condition characterised by chromosome complement 2n = 44 + XXY ₅
9) tetraploid	**i)** non fertile offspring from cross between two genetically unlike parents ₈
10) Turner's syndrome	**j)** large molecule composed of one or more chains of amino acids ₇

8 Figure 9.24 shows the karyotypes of two humans. Each possesses an unusual chromosome complement.
a) Summarise the chromosome complement in each case using numbers and the letters X and/or Y where appropriate.
b) State the sex of each person.
c) (i) Name the syndrome suffered by each person.
(ii) State TWO characteristics typical of each syndrome.
d) Briefly explain how the abnormality in person B's chromosome complement probably arose.

101

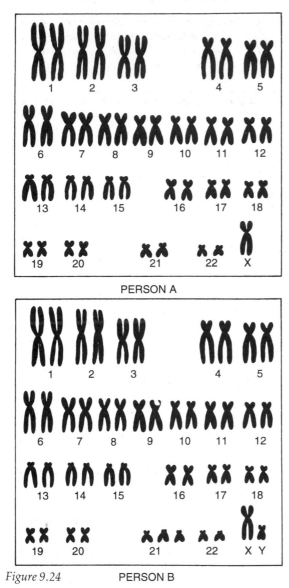

PERSON A

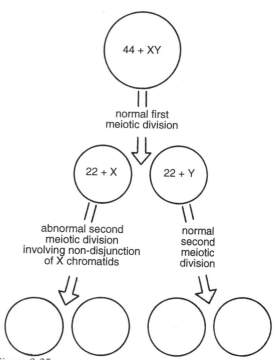

Figure 9.24 PERSON B

10 Figure 9.25 shows meiosis in a sperm mother cell. Although the first meiotic division is normal, one of the products is affected by non-disjunction during the second division.

Figure 9.25

a) Using only numbers and letters as required, represent the four sperm that would be formed.
b) Using the same convention, represent the zygote that would be formed if each of these sperm successfully fertilised a normal egg.
c) What percentage of the zygotes in your answer to b) would develop into sufferers of Turner's syndrome?

9 Rewrite the following sentences choosing the correct alternative from each choice

The sex of a person with Klinefelter's syndrome is $\begin{Bmatrix} \text{male.} \\ \text{female.} \end{Bmatrix}$ Such a person began life as a zygote containing the sex chromosomes $\begin{Bmatrix} \text{XXX} \\ \text{XXY} \end{Bmatrix}$ which could have resulted from an unusual $\begin{Bmatrix} \text{XY} \\ \text{XX} \end{Bmatrix}$ egg being fertilised by a normal $\begin{Bmatrix} \text{Y} \\ \text{X} \end{Bmatrix}$ sperm.

11 When the cells of a certain type of hyacinth plant are examined, each is found to contain 32 chromosomes as four sets all derived from the one species of plant. The hyacinth's chromosome complement can be represented as AAAA.
Rewrite the following sentences choosing only the correct option in each case.

The hyacinth plant described above is said to be $\begin{Bmatrix} \text{diploid} \\ \text{tetraploid} \end{Bmatrix}$ and could have arisen from the fusion of a $\left\{\begin{array}{l}\text{diploid}\\\text{haploid}\end{array}\right\}$ gamete with a $\left\{\begin{array}{l}\text{diploid}\\\text{haploid}\end{array}\right\}$ gamete. The basic haploid number of its ancestral plant would be $\left\{\begin{array}{l}8.\\16.\end{array}\right\}$

12 Table 9.1 refers to four sub-species of the celandine plant (*Ranunculus ficaria*).

sub-species of celandine	chromosome complement of normal plant cell	genomes present
1	16	AA
2	32	AAAA
3	40	AAAAA
4	48	AAAAAA

Table 9.1

a) What is meant by the term polyploid?
b) Which of the sub-species in the table would be described as polyploid?
c) What type of chromosomal change (mutation) gives rise to a polyploid?
d) Which sub-species in the table is sterile?
e) How many chromosomes are represented by the genome symbol A?
f) Suggest how sub-species 4 arose.

13 Table 9.2 refers to three species of the *Brassica* group of plants.
a) Copy and complete table 9.2.
b) Scientists consider cabbage and turnip to be the original parents of swede but a cross between cabbage and turnip produces a sterile hybrid.
(i) Represent this sterile hybrid in terms of genome letter symbols.
(ii) State the number of chromosomes that would be present in one of the sterile hybrid plant's body cells.
(iii) Construct a flow chart to show how a swede could have arisen from the sterile hybrid.

14 *Spartina townsendii* is a fertile polyploid grass plant which is thought to have arisen from two of its close relatives. Table 9.3 refers to the three species of plant.

scientific name	chromosome number present in each gamete	chromosome number present in each body cell	genomes present in each body cell
Spartina maritima			AA
Spartina alterniflora		70	BB
Spartina townsendii	63	126	AABB

Table 9.3

a) Copy and complete the table.
b) The formation of *S. townsendii* could have involved the following steps which are given in a mixed-up order. Arrange them in the correct sequence.
   1 complete non-disjunction of chromosomes in cell of sterile hybrid.
   2 vegetative growth of sterile hybrid
   3 growth of polyploid cell into fertile *S. townsendii* plant
   4 fertilisation of *S. maritima* by *S. alterniflora*
   5 formation of polyploid cell
   6 complete spindle failure during mitosis in cell of sterile hybrid
   7 formation of sterile hybrid

15 Figure 9.26 shows how one variety of edible banana plant arose.
a) Copy the diagram and complete all of the blank boxes.
b) Briefly explain why this variety of edible banana plant is sterile.

scientific name	common name	chromosome number present in each gamete	chromosome number present in each body cell	genomes present in each body cell
Brassica oleracea	cabbage		18	AA
Brassica rapa	turnip	10		BB
Brassica rapo-brassica	swede		38	AABB

Table 9.2

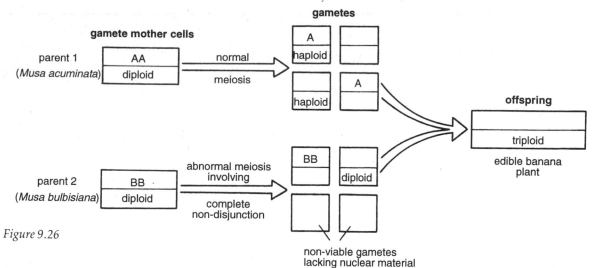

Figure 9.26

c) What process must occur to change this sterile plant into a fertile variety?

d) Another closely related triploid variety of edible banana has the genome combination AAB. Suggest how it arose.

16 Decide whether each of the following statements is true or false and then use T or F to indicate your choice. Where a statement is false, give the word(s) that should have been used in place of the word(s) in **bold print**.

a) Proteins are made up of many tiny sub-units called **amino acids**.

b) The sequence of a protein's sub-units is determined by a particular **genetic code** in a gene on a chromosome.

c) A change in structure of a gene leads to an alteration in the **chromosome** that the gene codes for.

d) Sufferers of sickle cell anaemia possess an unusual blood protein called **haemoglobin S**.

e) A molecule of normal haemoglobin differs from a molecule of haemoglobin S by one **protein**.

17 What name is given to a disturbance or abnormality in function of some part or system of the human body?

18 Look at table 9.4 which takes the form of a grid.

1 result of mutation affecting X chromosome	2 result of mutation affecting chromosome 7
3 production of abnormally thick mucus	4 production of blood-clotting factor VIII
5 blockage of pancreatic duct	6 severe congestion of lungs

Table 9.4

a) Choose ONE OR MORE statements from the grid which refer to cystic fibrosis.

b) Choose ONE OR MORE statements from the grid that refer to haemophilia.

19 Rewrite the following sentences and complete the blanks.

In the future it may be possible to replace defective genes with healthy ones. This form of treatment is called _____ therapy. In years to come, altered _____ cells may be able to make _____ _____ and prevent the blood of haemophiliacs from failing to _____ .

20 List the sequence of events which leads to the occurrence of a biochemical dysfunction (e.g. phenylketonuria) in the human body.

10 CONCEPTS OF GENETICS

MONOHYBRID CROSS

A cross between two true-breeding parents which differ in one way is called a **monohybrid cross**.

Gregor Mendel (1822–84), an Austrian monk, carried out early monohybrid crosses using varieties of pea plant. By appreciating the importance of working with **large numbers** of plants, studying **one characteristic** at a time and **counting** the offspring produced, Mendel was the first to put genetics on a firm scientific basis.

In the experiment shown in figure 10.1, Mendel crossed pea plants which were true-breeding for production of round seeds with pea plants true-breeding for wrinkled seeds.

All the seeds produced in the **first filial generation** (F_1) were round. Once these seeds had grown into plants, they were self-pollinated. The resultant **second filial generation** (F_2) consisted of 7324 seeds (5474 round and 1850 wrinkled). This is a ratio of 2.96 : 1. When these figures are analysed statistically they are found to represent a ratio of 3 round : 1 wrinkled.

Since wrinkled seed, absent in the F_1, reappears in the F_2, something has been transmitted undetected in the gametes from generation to generation. Mendel called this a factor. Today we call it a **gene**. In this case it is the gene for seed shape which has two forms (**alleles**): round and wrinkled.

Since the presence of the round allele masks the presence of the wrinkled allele, round is said to be **dominant** and wrinkled **recessive**.

Using symbols 'R' for round and 'r' for wrinkled, the cross can be summarised as follows:

Original cross	RR × rr
gametes	all R ↓ all r
F_1 genotype	all Rr
phenotype	all round
second cross	Rr × Rr (self pollinated)
gametes	R and R ↓ R and r

F_2 genotypes (in Punnett square)	ovules		R	r
		R	RR	Rr
		r	rR	rr

F_2 phenotypic ratio   3 round : 1 wrinkled

KEY QUESTIONS

1 What term is used to describe a cross between two true-breeding parents which differ from one another in a way only?

2 Give THREE examples of good scientific practice employed by Mendel in his experiments with pea plants.

3 **a)** Using symbols, represent the genotype of: (i) a plant which is true-breeding for production of round seeds; (ii) a plant which is true-breeding for production of wrinkled seeds; (iii) the F_1 offspring that result from crossing (i) with (ii).
b) State the phenotype of the F_1 offspring.
c) (i) When these F_1 offspring are self-pollinated, what F_2 phenotypic ratio is produced? (ii) Draw a Punnett square to show how this ratio arises.

4 **a)** What term is used to refer to the different expressions of a gene?
b) Which expression of the gene for seed shape was masked in the F_1 in Mendel's experiment but reappeared in the F_2?
c) What term is used to apply to an allele that can be masked by another allele?
d) Which allele of the gene for seed shape is dominant?

DIHYBRID CROSS

A cross between two true-breeding parents, which differ from one another in **two** ways, is called a **dihybrid cross**. In one of his experiments, Mendel crossed true-breeding pea plants which produced round yellow seeds with true-breeding plants which produced wrinkled green seeds.

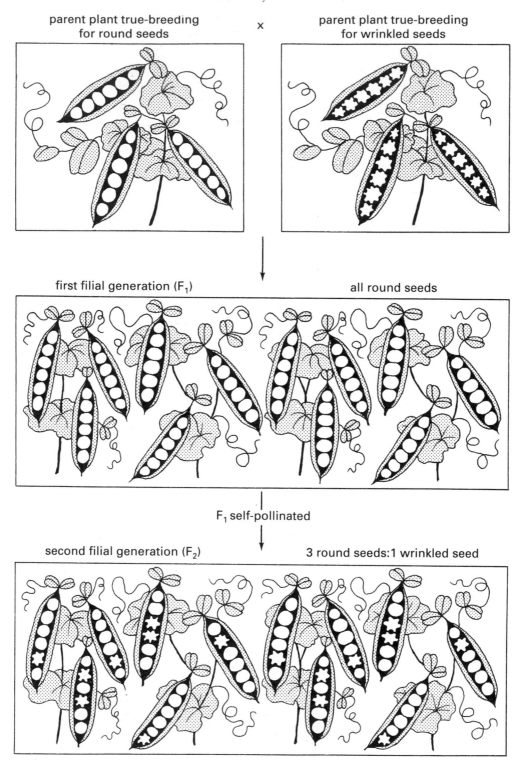

Figure 10.1:   Monohybrid cross

parent plant true-breeding
for round yellow seeds   ×   parent plant true-breeding
for wrinkled green seeds

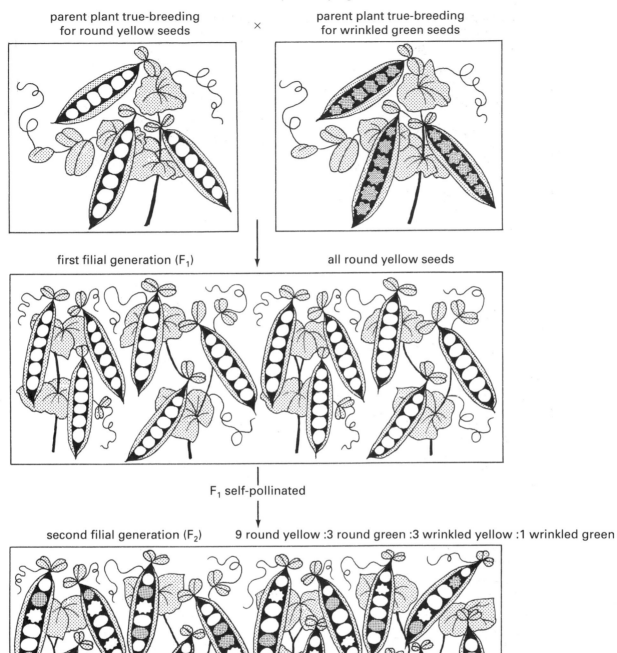

first filial generation (F₁)   all round yellow seeds

F₁ self-pollinated

second filial generation (F₂)   9 round yellow :3 round green :3 wrinkled yellow :1 wrinkled green

Figure 10.2:   Dihybrid cross

107

This cross involves two genes: the gene for seed shape (alleles – round and wrinkled) and the gene for seed colour (alleles – yellow and green).

All the F_1 plants produced round yellow seeds. This shows that round is dominant to wrinkled and yellow is dominant to green. This cross was followed through to the F_2 generation. A simplified version of it is shown in figure 10.2. The actual observed F_2 consisted of 315 round yellow seeds, 108 round green, 101 wrinkled yellow and 32 wrinkled green.

When these figures are analysed statistically they are found to represent a **9 : 3 : 3 : 1 ratio**. Figure 10.3 shows how an F_2 with such a phenotypic ratio can arise. For the sixteen possible combinations shown in the Punnett square to occur, each F_1 parent at the second cross must make four different types of gamete in equal numbers.

Let R = allele for round

and r = allele for wrinkled

Let Y = allele for yellow

and y = allele for green

original cross	RRYY × rryy
gametes	all RY ↓ all ry
F_1	all Rr Yy
second cross	RrYy × RrYy (F_1 self-fertilised)
gametes	RY, Ry ,rY, ry ↓ RY ,Ry, rY, ry

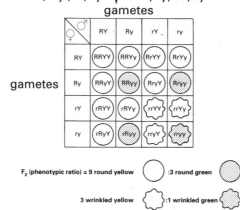

gametes

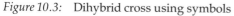

F_2 (phenotypic ratio) = 9 round yellow ⃝ :3 round green ⬤

3 wrinkled yellow :1 wrinkled green

Figure 10.3: Dihybrid cross using symbols

How does this occur? If the genes for seed shape and colour are located on different chromosomes then these will be subjected to **random (independent) assortment** during meiosis as

shown in figure 8.3 on page 78. There are two possible ways in which homologous pairs can become arranged at the equator. As a result, approximately 50% of gamete mother cells produce RY and ry gametes and the other 50% produce Ry and rY gametes.

This principle can be demonstrated by carrying out the bead model of gamete formation experiment shown in figure 10.4.

KEY QUESTIONS

1 What is meant by the expression *a dihybrid cross?*

2 **a)** State the phenotype of each of the true-breeding parents chosen by Mendel for the dihybrid cross described in the text.
b) State the phenotype of the F_1 generation resulting from crossing these two parents.
c) Which allele of the gene for seed shape must therefore be dominant?
d) Similarly which allele of the gene for seed colour must be dominant?

3 **a)** Using symbols represent the genotype of:
(i) a plant which is true-breeding for round yellow seeds; (ii) a plant which is true-breeding for wrinkled green seeds; (iii) the F_1 offspring resulting from a cross between (i) and (ii).
b) (i) When these F_1 offspring are self-pollinated, what F_2 phenotypic ratio is produced? (ii) Draw a Punnett square to show how this ratio arises.

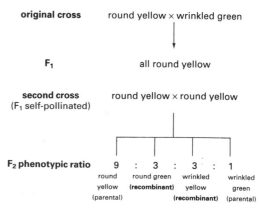

original cross	round yellow × wrinkled green
F_1	all round yellow
second cross (F_1 self-pollinated)	round yellow × round yellow

F_2 phenotypic ratio

9 : 3 : 3 : 1
round yellow (parental) : round green (recombinant) : wrinkled yellow (recombinant) : wrinkled green (parental)

Figure 10.5: Recombinants

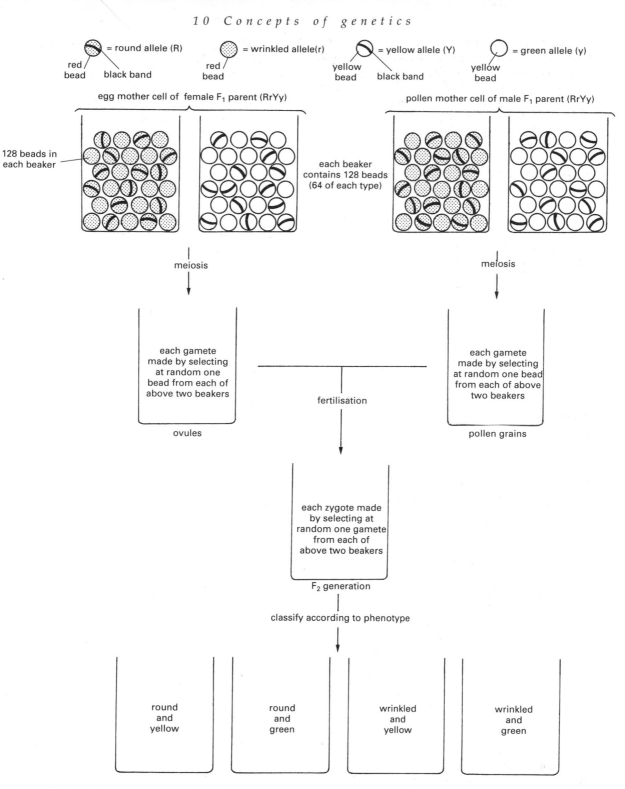

Figure 10.4: Bead model of gamete formation

RECOMBINATION

In a dihybrid cross, two of the F_2 phenotypes resemble the original parents and two display new combinations of the characteristics. This process by which new combinations of parental characteristics arise is called **recombination**. The individuals possessing the new combinations are called **recombinants** (see figure 10.5).

DIHYBRID BACKCROSS (TESTCROSS)

If the F_1 pea plants produced in the example shown in figure 10.3 are backcrossed (testcrossed) with a **double recessive** plant (instead of being self-fertilised), then the following occurs:

Second cross $\qquad$ RrYy × rryy

gametes $\quad$ RY, Ry, rY, ry $\quad \downarrow \quad$ all ry

F_2 genotypes
$\qquad$ RrYy $\qquad$ Rryy $\qquad$ rrYy $\qquad$ rryy

F_2 phenotypes
$\qquad$ round, $\quad$ round, $\quad$ wrinkled, $\quad$ wrinkled
$\qquad$ yellow $\quad$ green $\qquad$ yellow $\qquad$ green
genotypic and phenotypic ratio
$\qquad$ 1 : $\qquad$ 1 : $\qquad$ 1 : $\qquad$ 1

CONCLUSIONS

In a dihybrid cross where the genes show independent (random) assortment, **self-fertilising** the F_1 results in an F_2 with the phenotypic ratio **9 : 3 : 3 : 1** whereas **backcrossing** the F_1 results in an F_2 with the phenotypic (and genotypic) ratio **1 : 1 : 1 : 1**.

KEY QUESTIONS

1 a) How many different phenotypes occur in the F_2 generation of a dihybrid cross involving two genes which show random assortment?
b) Of these different F_2 phenotypes, how many types resemble the original parents and how many types display new combinations of the characteristics?

2 a) What name is given to the process by which new combinations of characteristics arise?

b) What term is used to refer to the individuals possessing the new combinations of characteristics?

3 a) What F_2 phenotypic *ratio* results from a dihybrid cross (involving two genes showing random assortment) following self-fertilisation of the F_1?
b) What *proportion* of this F_2 generation could be described as showing (i) both parental characteristics? (ii) recombinant characteristics?

4 a) What F_2 phenotypic *ratio* results from a dihybrid cross (involving two genes showing random assortment) following a backcross of the F_1 to double recessive individuals?
b) What *proportion* of this F_2 generation could be described as showing (i) both parental characteristics? (ii) recombinant characteristics?

LINKED GENES

Since an organism possesses many more genes than chromosomes, it follows that each chromosome must carry **many** genes. Genes on the same chromosome do not behave independently of one another. When a cross involves two alleles of two different genes located on the same chromosome, the alleles of the two genes are **transmitted together** at meiosis and are said to be **linked**.

Consider the following example for tomato plants involving two linked genes, the gene for height (alleles T for tall, t for dwarf) and the gene for stem type (alleles S for smooth and s for hairy). If the two genes were completely linked then a backcross (testcross) would fail to produce any F_2 recombinants as shown in figure 10.6.

However when this cross is carried out, the F_2 is found to contain a **few** tall hairy and a **few** dwarf smooth recombinants. So how did these recombinants arise?

MECHANISM BY WHICH LINKED GENES ARE SEPARATED

During meiosis when homologous chromosomes form pairs, **crossing over** may occur between adjacent chromatids at points called chiasmata (also see page 79). If crossing over occurs between two genes, this separates alleles that were previously linked and allows them to recombine in new combinations.

Figure 10.7 on p.112 shows how the tall hairy (Ttss) and dwarf smooth (ttSs) recombinants arise.

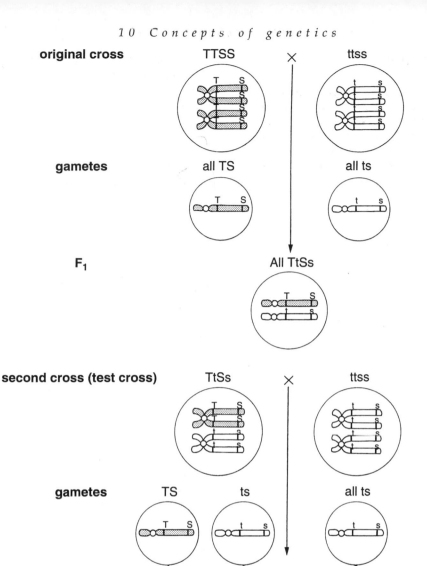

Figure 10.6: Complete linkage

Conclusion

If a dihybrid backcross between the F_1 and the double recessive parent **fails** to produce an F_2 in a ratio of 1 : 1 : 1 : 1 then this indicates that the two genes involved are **linked**.

Similarly if a dihybrid cross involving self-fertilisation of the F_1 **fails** to produce an F_2 in a ratio of 9 : 3 : 3 : 1, this also indicates that the two genes involved are **linked**.

KEY QUESTIONS

1 **a)** Why is it not possible that each chromosome in an organism carries only one gene?

   b) What are linked genes?

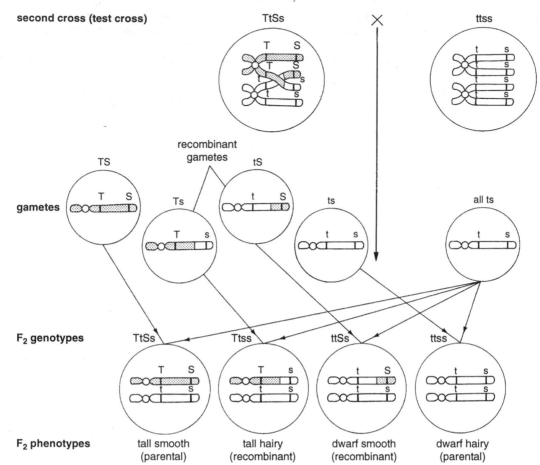

second cross (test cross) TtSs × ttss

recombinant
gametes

gametes TS Ts tS ts all ts

F₂ genotypes TtSs Ttss ttSs ttss

F₂ phenotypes tall smooth (parental) tall hairy (recombinant) dwarf smooth (recombinant) dwarf hairy (parental)

Figure 10.7: Separation of linked genes

2 a) Referring to the example in the text, state the genotype of: (i) a true-breeding tall smooth-stemmed tomato plant; (ii) a true-breeding dwarf hairy-stemmed tomato plant; (iii) the F_1 generation resulting from a cross between (i) and (ii).
b) State the genotypes of the F_2 generation that would be formed by backcrossing (testcrossing) this F_1 if the two genes were completely linked.
c) In reality a few members of the F_2 have the genotype Ttss and ttSs. (i) What general name is given to such non-parental forms? (ii) Copy and complete the following sentence. In addition to making TS and ts gametes, F_1 plants (genotype TsSs) also produce a few ____ and ____ gametes. These recombinant gametes occur as a result of ____ ____ occurring between adjacent chromatids during ____ .

3 a) What F_2 phenotypic ratio results from backcrossing (testcrossing) a dihybrid F_1 (e.g. AaBb) where the two genes involved are on different chromosomes?
b) What F_2 phenotypic ratio would result from self-fertilising this dihybrid F_1 (AaBb)?
c) Neither of these ratios was obtained when F_1 dihybrid CcDd was both backcrossed and self-fertilised. What does this indicate about genes C/c and D/d?

SEX RATIOS AND SEX DETERMINATION

There are 46 chromosomes in the nucleus of every normal human body cell. These exist as 22 homologous pairs of chromosomes which play no part in sex determination and one pair of **sex chromosomes** which determine an individual's sex.

In the female, the sex chromosomes make up a homologous pair, the X chromosomes. Thus a human female has the chromosome complement **44 + XX** and every egg formed contains an X chromosome.

In the male, the sex chromosomes make up an unmatched but homologous pair consisting of an X chromosome and a smaller Y chromosome. Thus a human male has the chromosome complement **44 + XY.** Half of the sperm formed contain an X chromosome and the other half a Y chromosome.

When the nucleus of a sperm fuses with the nucleus of an egg at fertilisation, the sex of the zygote is determined by the type of sex chromosome carried by the sperm. Approximately 50 per cent of zygotes are XX (female) and 50 per cent are XY (male) giving a **sex ratio of 1 : 1** as shown in figure 10.8.

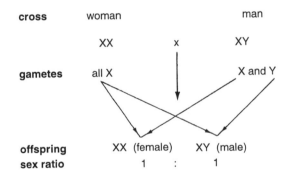

Figure 10.8: Sex ratio

KEY QUESTIONS

1 a) State the total number of chromosomes present in the nucleus of a normal human body cell.
b) How many of these are sex chromosomes?
c) Use letters to represent the sex chromosomes possessed by a normal human (i) male (ii) female.

2 a) Referring only to sex chromosomes, state how many types of (i) egg are made by a normal human female (ii) sperm are made by a normal human male.

b) Draw a simple diagram to show how male and female zygotes are formed as a result of fertilisation.
c) By which type of sex cell is the sex of a zygote determined?
d) State the sex ratio of zygotes that occurs when a large human population is considered.

SEX LINKAGE

Although sex chromosomes behave as a homologous pair at meiosis, an X and a Y chromosome do not make up a truly homologous pair. In humans for example, the larger X chromosome carries many genes that are not present in the smaller Y chromosome. This is shown in figure 10.9.

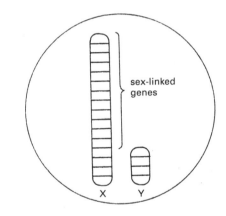

Figure 10.9: Sex-linked genes

These genes are said to be **sex-linked.** When a sex-linked gene occurs on the X chromosome but not on the Y, and that X chromosome meets a Y at fertilisation, then the sex-linked characteristic (whether dominant or recessive) will be expressed in the phenotype of the organism produced. This is because the Y chromosome has no allele at the equivalent gene locus to offer dominance.

SEX LINKAGE IN FRUIT FLY

Fruit flies have the same mechanism of sex determination as humans. Although the Y chromosome is not smaller than the X, it is a different shape and carries very few genes. The X chromosome therefore carries many sex-linked genes. One of these determines eye colour.

The allele for red eye colour (R) is dominant to the allele for white eye (r). Figure 10.10 shows a cross between a white-eyed female and a red-eyed male followed through to the F₂ generation.

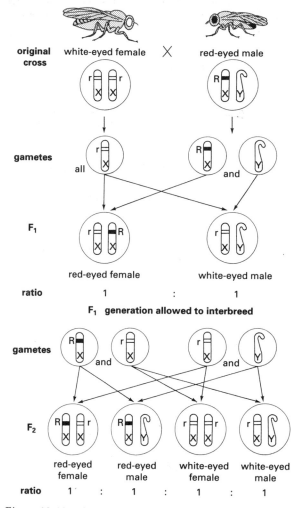

Figure 10.10: Sex linkage in fruit fly

A simpler version of this cross is shown in figure 10.11 where the sex chromosomes are represented by X and Y and the alleles of the sex-linked gene by the superscripts of R and r.

Since this cross involves a sex-linked gene, the F₂ generation does not show the phenotypic ratio of 3 : 1 typical of a normal monohybrid cross.

SEX LINKAGE IN HUMANS
Red-green colour-blindness

In humans, normal colour vision (C) is dominant to **red-green colour-blindness** (c). These are the

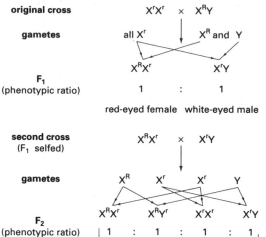

Figure 10.11: Sex linkage in fruit fly using symbols

genotype	phenotype
$X^C X^C$	female with normal colour vision
$X^C X^c$	female (carrier) with normal colour vision
$X^c X^c$	female with colour-blindness (very rare e.g. 0.5% of European population)
$X^C Y$	male with normal colour vision
$X^c Y$	male with colour-blindness (more common e.g. 8% of European population)

Table 10.1   Red-green colour-blindness in humans

alleles of a sex-linked gene on the X chromosome. The five possible genotypes are given in table 10.1.

Heterozygous females are **carriers** because, although unaffected themselves, they pass the allele on to 50 per cent of their offspring. On average, therefore, 50 per cent of a carrier female's sons are colour-blind. This is shown in figure 10.12.

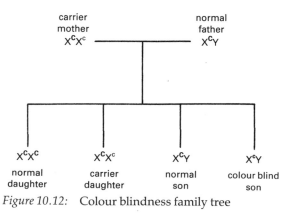

Figure 10.12: Colour blindness family tree

Red-green colour-blindness is rare in females since two recessive alleles must be inherited. It is more common in males where only one is needed.

Haemophilia

Haemophilia is a disorder involving defective blood clotting. It is caused by a recessive allele carried on the X chromosome and is therefore sex-linked.

KEY QUESTIONS

1 a) Which of the human sex chromosomes is the larger one?
b) What term is used to refer to genes which are present on the X chromosome but absent from the Y chromosome?
c) If a human male inherits the recessive allele of a sex-linked gene, it is always expressed in his phenotype. Explain why.

2 a) Using symbols, represent the genotype(s) of:
(i) a true-breeding white-eyed female fruit fly;
(ii) a true-breeding red-eyed male fruit fly;
(iii) the F_1 offspring that would result from a cross between (i) and (ii).
b) Draw a Punnett square to show the outcome of crossing F_1 males with F_1 females.
c) State the phenotypic and sex ratios of the F_2 offspring.
d) What would have been the phenotypic ratio of the F_2 offspring, had eye colour not been a sex-linked gene?

3 a) Represent symbolically the genotype(s) of: (i) a colour-blind human male; (ii) a human female with normal vision who is a carrier; (iii) the offspring that could be produced if (i) and (ii) were their parents.
b) Name a sex-linked condition which involves defective blood clotting in humans.
c) Why are such sex-linked conditions expressed much less frequently in the phenotype of females compared with males?

MULTIPLE ALLELES

In the monohybrid cross described on page 102, the gene for seed shape was found to have only two alleles (round and wrinkled). However this situation does not apply to all genes. In some cases **several alleles** of a gene exist and are available to occupy that particular gene site (**locus**) on a chromosome. Obviously each diploid individual can only possess a maximum of two of these alleles (one from the male parent and one from the female parent).

RABBIT COAT COLOUR

Within a population of domestic rabbits, four alleles of the gene for coat colour exist. These show complete dominance in the order C (full colour) dominant to C^{ch} (chinchilla) dominant to C^H (Himalayan) dominant to C^a (albino).

The phenotype of each homozygous genotype is shown in figure 10.13. Table 10.2 gives the phenotypes of the heterozygous genotypes.

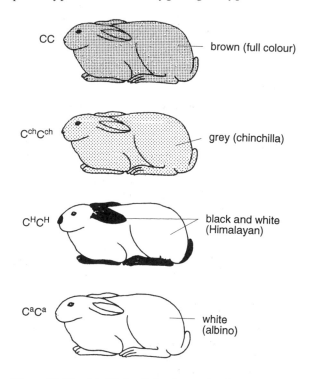

Figure 10.13: Multiple alleles for coat colour

heterozygous genotype(s)	phenotype (coat colour)
CC^{ch}, CC^H, CC^a	brown (full colour)
C^{ch} C^H, C^{ch} C^a	grey (chinchilla)
C^H C^a	black and white (Himalayan)

Table 10.2   Multiple alleles for coat colour

HUMAN BLOOD GROUP

The gene for blood group in humans has three alleles. A and B are co-dominant to recessive O (see table 10.3).

genotype(s)	phenotype (blood group)
AA, AO	A
BB, BO	B
AB	AB
OO	O

Table 10.3   Multiple alleles for blood group

KEY QUESTIONS

1 **a)** What term is used to refer to the particular site on a chromosome at which a gene is located?
b) What name is given to the three or more forms of a gene that are able to occupy the same locus on a chromosome?
c) What is the maximum number of different alleles of a gene that can be possessed by an individual diploid organism? Explain why.

2 **a)** Using symbols, give the genotype(s) of: (i) an albino rabbit; (ii) a true-breeding full colour rabbit; (iii) a true-breeding chinchilla rabbit; (iv) a heterozygous Himalayan rabbit.
b) State the colour of fur that would be possessed by rabbits with the following genotypes: (i) CC^H (ii) C^aC^a (iii) $C^{ch}C^a$ (iv) C^HC^a.

3 **a)** Using symbols, give the genotype(s) of blood groups: (i) A (ii) B (iii) AB.
b) Draw a Punnett square to show the outcome of a cross between parents with genotypes AO and BO.

INCOMPLETE DOMINANCE

In some cases it is found that one allele of a gene is **not** completely dominant over the other allele.

FLOWER COLOUR IN SNAPDRAGON PLANTS

When true-breeding red snapdragon plants are crossed with true-breeding white (ivory) plants, the F_1 are all **pink.** When the F_1 is self-pollinated,

the F_2 occurs in the ratio 1 homozygous red: 2 heterozygous pink: 1 homozygous ivory. This cross is shown in figure 10.14 where R = allele for red flower and I = allele for ivory flower.

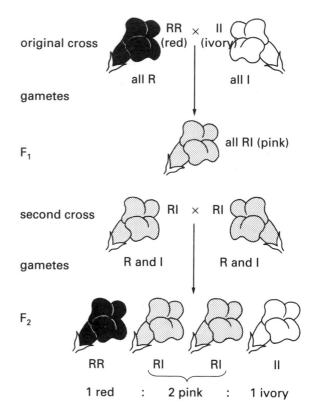

Figure 10.14:   Incomplete dominance in snap dragon

COAT COLOUR IN SHORTHORN CATTLE

The alleles of the gene for coat colour in shorthorn cattle are red (R) and white (W). They are incompletely dominant to one another. The heterozygote (RW) inherits a mixture of red and white hairs in its coat which is described as **roan** in colour. The result of a cross between a roan bull and a roan cow is shown in figure 10.15.

In all cases of incomplete dominance, heterozygous individuals are phenotypically different from both of their homozygous parents.

KEY QUESTIONS

1 What term is used to describe the relationship between two alleles where neither one is able to

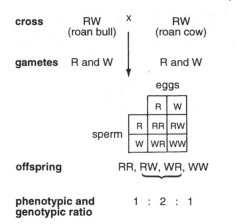

Figure 10.15: Cross between two roan cattle

mask completely the phenotypic expression of the other?

2 **a)** Using symbols, represent the genotype of a red snapdragon plant.
b) Why are the symbols II and not rr used to represent an ivory plant?
c) What phenotype is possessed by all members of the F_1 generation resulting from a cross between a red and an ivory plant?
d) Draw up a Punnett square to show the outcome of self-pollinating pink snapdragons.

3 **a)** Using symbols, represent the genotype of:
(i) a red-coated shorthorn cow; (ii) a white-coated shorthorn bull; (iii) the F_1 offspring that would result from a cross between (i) and (ii).
b) Give the phenotypic and genotypic ratios of the F_2 generation resulting from crosses between F_2 roan cattle.

EXERCISES

1 Match the terms in list **X** with their descriptions in list **Y**.

list X	list Y
1) genotype	a) an organism's appearance resulting from genetic information inherited from parents
2) phenotype	b) genotype possessing two different alleles of a gene
3) dominant allele	c) the set of genes possessed by an organism
4) recessive allele	d) the offspring resulting from a cross
5) hetero-zygous	e) form of a gene that masks the presence of the recessive form
6) homo-zygous	f) genes located together on the same chromosome
7) linked genes	g) form of a gene that is masked by the presence of the dominant form
8) progeny	h) genotype possessing two identical alleles of a gene

2 In maize plants, the genes for leaf type and plant height are located on different chromosomes. Smooth leaf (S) is dominant to crinkly leaf (s) and tall stem (T) is dominant to short (t).
A plant heterozygous for both of these genes was self-pollinated.
a) Draw up a Punnett square to show the genotypes of the gametes involved and the offspring that would result.
b) State the phenotypes of the offspring and the ratio in which they would occur.
c) Give the genotypes and phenotypes of two of the offspring in your Punnett square which, if crossed, could produce progeny with four phenotypes in equal numbers.

3 In pea plants, tall is dominant to dwarf and red flower colour is dominant to white. A tall plant with red flowers was crossed with a dwarf plant with white flowers. The offspring were as follows: 47 tall red: 43 tall white: 46 dwarf red: 44 dwarf white.
a) Express this phenotypic ratio in its simplest form (taking into account that in genetics experiments, actual results normally vary slightly from expected results).
b) (i) Choose suitable symbols to represent the two alleles of each gene and state these clearly. (ii) Using your symbols, show the above cross in diagrammatic form to include the genotypes of both parents and all types of offspring.

4 In tomato plants, purple stem (P) is dominant to green stem (p) and tall (T) is dominant to dwarf (t). The genes for stem colour and plant height are located on different chromosomes.
A plant heterozygous for both of these genes was self-pollinated and produced 160 seeds. Draw a Punnett square to enable you to calculate the expected number of progeny that would have:
a) the genotype ppTt;
b) the phenotype purple stem, dwarf height;

	black fur, yellow tips	black fur, no yellow tips	brown fur, yellow tips	brown fur, no yellow tips
parents	1			1
F_1	5♂, 5♀			
F_2 family 1	4	0	2	2
2	5	3	1	0
3	3	2	3	1
4	6	1	0	0
5	2	3	2	0
6	7	0	1	0
7	6	2	1	0
8	4	1	3	1
9	5	2	0	0
10	3	1	2	1
F_2 totals				

Table 10.4

c) the most common genotype;
d) the least numerous phenotype.

5 In mice, black fur (B) is dominant to brown (b) and the presence of yellow tips amongst fur (Y) is dominant to absence of yellow tips (y).

Table 10.4 gives the results from a series of breeding experiments. Each F_1 female produced two separate families as a result of being mated twice with one of the males.
a) Express the sex ratio of the F_1 generation in its simplest form.
b) State the genotype of the F_1 mice.
c) Calculate the F_2 totals and then express the F_2 phenotypic ratio in its simplest form.
d) (i) If the F_2 ratio had been 13 : 1 : 1 : 5, what would this have told you about the location of the two genes involved in this cross? (ii) In the light of your answer to part c), what conclusion can you draw about the location of the two genes?

6 In tomato plants, round fruit shape is dominant to pear shape and red fruit colour is dominant to yellow.
a) In diagrammatic form, follow to the F_2 generation a cross between a parent bearing pear-shaped yellow fruit and one which is true-breeding for round red fruit. (Assume that the F_1 is self-pollinated and that the two genes are located on different chromosomes.)
b) In your Punnett square, underline FOUR individuals possessing different genotypes and phenotypes.

7 The four fruit flies in figure 10.16 are all offspring from the same parents.
a) Fly A is homozygous recessive for the alleles

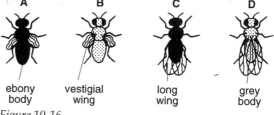

ebony body · vestigial wing · long wing · grey body

Figure 10.16

of the genes affecting wing type and body colour. Give the fly a genotype.
b) Assume that in the cross, one parent had the same genotype as A. What gametes must this parent have made?
c) What types of gamete must have been produced by the second parent?
d) State the genotypes of the second parent and of each of flies B, C and D.

Exercises 8, 9 and 10 are multiple choice items. In each case you should choose ONE correct answer only.

8 In fruit fly, dumpy wing (d) is recessive to normal wing (D) and ebony body (e) is recessive to normal body colour (E). The two genes involved are not on the same chromosome. A true-breeding normal-winged, ebony-bodied fly is crossed with a true-breeding dumpy-winged, normal-bodied fly.
a) The genotype of the F_1 generation will be
A Dd/Ee.   B DD/Ee.   C DD/EE.   D Dd/EE.
b) As a result of interbreeding amongst the members of the F_1 generation, dumpy-winged, normal-bodied flies will be present in the F_2 generation in the proportion
A 1 in 16.   B 3 in 16.   C 6 in 16.   D 9 in 16.

9 In mice, black coat (allele B) is dominant to white coat (b) and straight whiskers (allele S) is dominant to curved whiskers (s).

When true-breeding mice with black coats and straight whiskers were crossed with white mice possessing curved whiskers, the offspring were all black with straight whiskers.

If these F_1 mice were crossed with white mice possessing curved whiskers, the expected proportion of offspring with black coats and curved whiskers in the next generation would be

A 1 in 16.   B 3 in 16.   C 4 in 16.   D 9 in 16.

10 When a true-breeding plant bearing disc-shaped yellow fruit was crossed with a plant bearing sphere-shaped green fruit, the F_1 generation were all found to bear disc-shaped yellow fruit.

This F_1 generation was self-pollinated and produced 800 offspring. Predict which of the proportions of phenotypes shown in table 10.5 was found in this F_2 generation.

	phenotype			
	disc-shaped yellow	disc-shaped green	sphere-shaped yellow	sphere-shaped green
A	199	201	202	198
B	49	153	152	446
C	800	0	0	0
D	453	148	147	52

Table 10.5

11 In maize plants, white pericarp (W) is dominant to brown (w) and full endosperm (F) is dominant to shrunken (f).

A plant breeder carried out the cross WwFf × wwff. She obtained 200 offspring and counted their four phenotypes carefully. Table 10.6 compares the results that she expected to get with those that she actually obtained.

	number of offspring	
phenotype	expected	actual
white pericarp, full endosperm	50	81
white pericarp, shrunken endosperm	50	14
brown pericarp, full endosperm	50	16
brown pericarp, shrunken endosperm	50	89

Table 10.6

a) Account for the difference between the expected and the actual results.
b) (i) Which phenotypes were the recombinants?
(ii) These recombinants types arose as a result of:
A random assortment;   B linkage;
C crossing over;   D incomplete dominance.
(Choose ONE correct answer only.)

12 In mice, brown coat colour (B) is dominant to albino (b) and normal movement (M) is dominant to waltzer movement (m). These two genes are located on the same chromosome.

A cross between mice which are homozygous for the alleles of these genes is shown in figure 10.17.

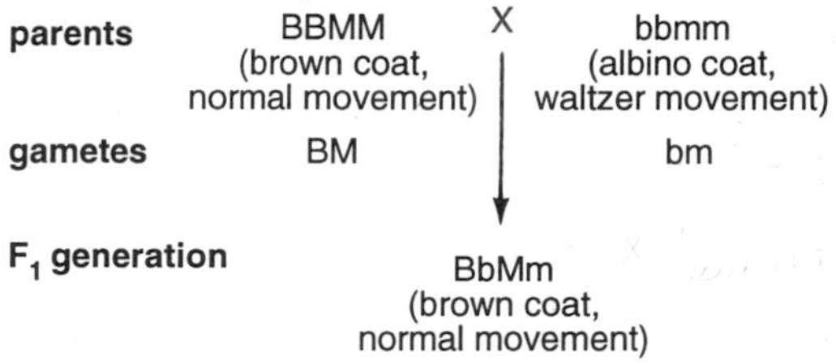

parents	BBMM (brown coat, normal movement)	X	bbmm (albino coat, waltzer movement)
gametes	BM		bm

F_1 generation   BbMm (brown coat, normal movement)

Figure 10.17

Mice from this F_1 generation were then crossed with mice homozygous for both recessive alleles.
a) Assuming that no crossing over occurred, state the following:
(i) F_2 genotypes;
(ii) F_2 genotypic ratios;
(iii) F_2 phenotypes;
(iv) F_2 phenotypic ratio.
b) If crossing over did occur between the two genes then two further F_2 genotypes would have been produced. State them.

13 Match the terms in list **X** with their descriptions in list **Y**

list X	list Y
1) sex ratio	a) female who is heterozygous for a sex-linked gene
2) X chromosome	b) gene which is present on one sex chromosome (e.g. X) but not on other (e.g. Y)
3) Y chromosome	c) smaller of the two human sex chromosomes
4) sex-linked gene	d) three or more alleles of a gene that can occupy the same locus on the chromosome

5) sex linkage

6) carrier

7) multiple alleles

8) incomplete dominance

e) relative number of males and females in a population (normally 1 : 1 approximately)

f) condition of a heterozygote in which phenotype is different from both homozygous parental forms

g) larger of the two human sex chromosomes

h) inheritance pattern of alleles of genes carried on one sex chromosome but not on the other

14 a) Copy and complete figure 10.18 to illustrate the inheritance of sex chromosomes in cattle.

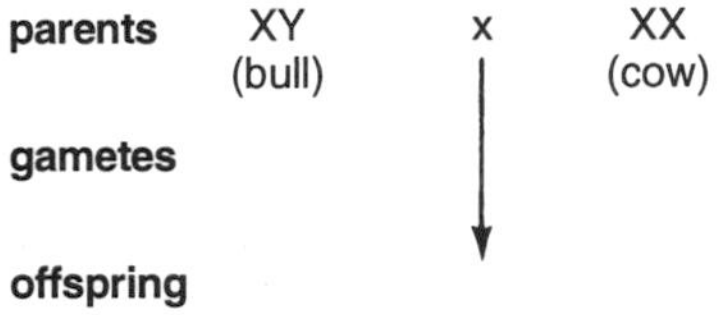

parents   XY   x   XX
            (bull)   (cow)

gametes

offspring

Figure 10.18

b) On your diagram state the sex ratio expected amongst the offspring.

15 In fruit flies, red/white eye colour is determined by a sex-linked gene where red (R) is dominant to white (r).
a) Using the symbols X^R, X^r and Y, draw a diagram to show the F_1 generation that would result from a cross between a white-eyed male and a homozygous red-eyed female.
Clearly state the expected genotypes, phenotypes and sex ratio of the F_1 offspring in your answer.
b) Show diagrammatically the F_2 generation that would result if members of the F_1 generation in part a) were allowed to interbreed.
Clearly state the expected genotypes, phenotypes and sex ratio of the F_2 generation in your answer.
c) What would have been the phenotypic and sex ratios of the F_2 generation had eye colour not been sex-linked?

16 In fruit flies, the gene for eye shape is sex-linked where bar eye (B) is dominant to normal round shape (b).

Draw a diagram to illustrate the outcome of a cross between a bar-eyed male and a round-eyed female. State the expected genotypes, phenotypes and sex ratio of the offspring in your answer.

17 Answer TRUE or FALSE to each of the following statements:
a) A man can pass on a sex-linked gene to his sons.
b) Red-green colour-blindness in humans is rarer in the female than in the male.
c) Females heterozygous for a sex-linked gene are called carriers.
d) The sex-linked gene for haemophilia is present on a Y but not on an X chromosome.
e) The sex of a human zygote is determined by the type of sex chromosome in the sperm.

18 Using the same format as figure 10.12, show the possible results of a cross between a colour-blind male and a carrier female.

19 Give the genotypes of the parents in a family where all the sons are colour-blind and all the daughters are normal-sighted carriers.

20 Haemophilia occurs in the family tree shown in figure 10.19.

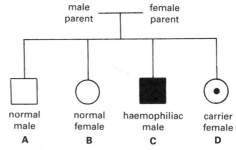

Figure 10.19

a) Using the convention X^H (normal allele), X^h (haemophiliac allele) and Y (no allele), give the genotypes of the offspring A, B, C and D.
b) Give the genotype and phenotype of each parent.
c) If C marries a normal female, what proportion of their sons are likely to be haemophiliacs?

21 In rabbits, homozygous combinations of four alleles of the gene for coat colour produce the phenotypes:
CC = full colour (brown)
$C^{ch}C^{ch}$ = chinchilla (grey)

C^HC^H = Himalayan (black and white)
C^aC^a = albino (white)
The alleles show complete dominance in the order $C>C^{ch}>C^H>C^a$
a) Stating genotypes and phenotypic ratios, show diagrammatically the expected outcome from each of the following crosses:
(i) $C^{ch}C^H \times CC^a$
(ii) $C^{ch}C^H \times C^{ch}C^H$
(iii) $CC^a \times CC^a$.
b) Table 10.7 shows the outcome of two crosses. Work out the genotypes of the parents in each cross.

22 A man with blood group A whose father was blood group O marries a woman with blood group AB.
a) Draw a diagram to show all the possible genotypes that could occur amongst their children.
b) Which phenotype could not occur amongst their offspring?

23 In the family tree shown in figure 10.20, a circle represents a woman and a square a man. The upper half of each circle or square contains the person's blood group phenotype and the lower half the person's blood group genotype. Some letters have been omitted.
Copy and complete figure 10.20.

24 In snapdragon plants, the alleles of the gene for flower colour, red (R) and ivory (I), are incompletely dominant to one another. Heterozygotes are found to be pink.
From the information in table 10.8, work out the phenotypes of the two parents used in each of crosses b–f.

25 In shorthorn cattle, the alleles of the gene for coat colour are red (R) and white (W) and they show incomplete dominance. Heterozygotes possess coats consisting of both red and white hairs and are described as roan in colour.

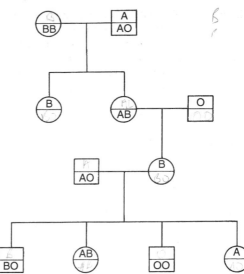

Figure 10.20

cross	phenotypes of two parents used	phenotypes and numbers of offspring produced
a)	red × white	158 pink
b)		41 red, 79 pink, 38 ivory
c)		161 red
d)		157 ivory
e)		81 red, 83 pink
f)		78 ivory, 82 pink

Table 10.8

a) Draw a diagram to show the possible genotypes and phenotypes amongst the offspring that would result from a cross between: (i) a red bull and a roan cow; (ii) a roan bull and a white cow.
b) If a roan bull and a roan cow were crossed repeatedly, approximately what proportion of their total offspring would also be roan?

cross	phenotypes of parents	phenotypes of offspring			
		full colour	chinchilla	Himalayan	albino
(i)	full colour × albino	4	0	5	0
(ii)	chinchilla × Himalayan	0	6	3	3

Table 10.7

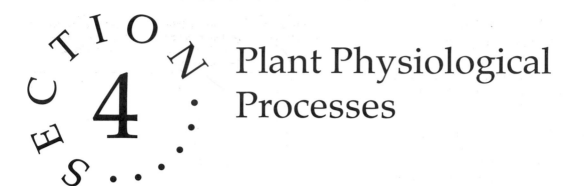

SECTION 4: Plant Physiological Processes

11 VITAL PHYSIOLOGICAL PROCESSES

PHOTOSYNTHESIS

Photosynthesis is the process by which green plants make **high-energy foods** (e.g. sugar and starch) from carbon dioxide and water using **light energy** trapped by green chlorophyll. Thus, during photosynthesis, green leaves convert light energy to chemical energy (contained in food). The process is summarised by the following word equation:

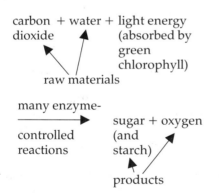

carbon + water + light energy
dioxide   (absorbed by
                        green
                        chlorophyll)
              raw materials

many enzyme-
controlled   sugar + oxygen
reactions   (and
                    starch)
              products

IMPORTANCE OF PHOTOSYNTHESIS

The **carbohydrate food** (e.g. sugar) and the **oxygen** produced during photosynthesis are used up during respiration in living cells to release **energy.** This is needed for vital processes such as growth, reproduction and cellular work (see figure 11.1).

Thus the survival of green plants (and all other life forms which feed directly or indirectly on green plants) depends on photosynthesis.

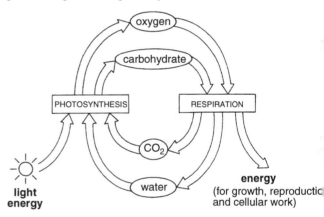

Figure 11.1: Importance of photosynthesis

KEY QUESTIONS

1 a) What is the function of chlorophyll?
   b) Name TWO raw materials used up during photosynthesis.
   c) Name TWO products of photosynthesis.

2 Give a simple version of the word equation of photosynthesis.

3 Explain why photosynthesis is important to the survival of a) green plants b) almost all other forms of life on Earth.

FACTORS AFFECTING PHOTOSYNTHETIC RATE

Several environmental factors affect the rate of photosynthesis. These include light intensity, carbon dioxide concentration, temperature and wavelength of light.

The **rate** of photosynthesis can be estimated by measuring one of the following:

- **evolution of oxygen** per unit time,
- **uptake of carbon dioxide** per unit time,
- **production of carbohydrate** (as increase in dry weight) per unit time.

INVESTIGATING THE EFFECT OF VARYING LIGHT INTENSITY

Elodea **bubbler experiment**

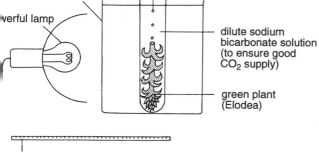

Figure 11.2: *Elodea* bubbler experiment

The number of **oxygen bubbles** released per minute by the cut end of an *Elodea* stem indicates the rate at which photosynthesis is proceeding. At first the lamp (see figure 11.2) is placed exactly 100 cm from the plant and the number of oxygen bubbles released per minute counted. The lamp is then moved to a new position (say 60 cm from the plant) and the rate of bubbling noted (once the plant has had a short time to become acclimatised to this new higher light intensity).

The process is repeated for lamp positions even nearer the plant as shown in table 11.1 When this typical set of results is displayed as a graph (figure 11.3), it can be seen that as light intensity increases, photosynthetic rate also increases until it reaches a maximum of 25 bubbles per minute at around 64 units of light.

distance from plant (cm)	units of light (calculated using mathematical formula)	number of oxygen bubbles released per minute
100	4	4
60	11	10
40	25	19
30	45	24
25	64	25
20	100	25

Table 11.1   Elodea bubbler results

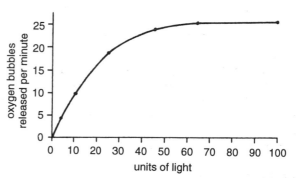

Figure 11.3:   Graph of *Elodea* bubbler results

Limiting factors

Further increase in light intensity does not increase the photosynthetic rate in the above experiment. This is because shortage of some other factor (e.g. amount of carbon dioxide) is now holding up the process. This factor in short supply is called a **limiting factor.**

INVESTIGATING THE EFFECT OF VARYING CARBON DIOXIDE CONCENTRATION

In this experiment the concentration of carbon dioxide made available to the *Elodea* plant is gradually increased by adding appropriate amounts of sodium bicarbonate to the water. The number of oxygen bubbles released is counted as before. The lamp is kept in one position to give uniform light of medium intensity.

The graph of a set of results (see figure 11.4) shows that when the plant is supplied with a carbon dioxide concentration of only 1 unit, photosynthesis is **limited** by this low concentration of carbon dioxide to 3 oxygen bubbles per minute. When carbon dioxide concentration is increased to 2 units, photosynthesis increases to 6 bubbles per minute

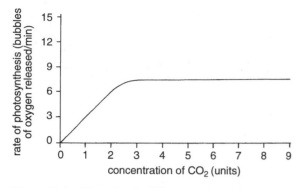

Figure 11.4: Variation in CO_2 concentration at one light intensity

but no further since carbon dioxide concentration becomes limiting again.

A further increase in carbon dioxide concentration to 3 units brings about a further increase in photosynthetic rate. However beyond this point the graph levels out and any further increase in carbon dioxide concentration does not affect photosynthetic rate. This is because some other factor (e.g. light intensity) is now limiting the process.

Effect of varying carbon dioxide concentration at different light intensities

In figure 11.5, graph ABC represents a repeat of figure 11.4 and graph ADE represents the results from a further *Elodea* bubbler experiment using the same plant in conditions of constant high light intensity.

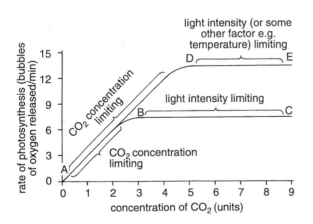

Figure 11.5: Variation in CO_2 concentration at two light intensities

In this further experiment an increase in carbon dioxide concentration to 4 and to 5 units brings about a corresponding increase in photosynthetic rate in each case. This is because when light intensity is higher, carbon dioxide concentration is still the **limiting factor** up to 5 units of carbon dioxide.

However beyond 5 units of carbon dioxide, the graph levels off again since some other factor (such as light intensity or temperature) has become limiting.

INVESTIGATING THE EFFECT OF VARYING TEMPERATURE

The apparatus shown in figure 11.2 is adapted for this experiment by exchanging the large trough for a water bath whose temperature is under thermostatic control. Plastic bags containing ice cubes are used to create low temperatures.

The plant is given light of constant high intensity and a rich supply of carbon dioxide to ensure that neither of these factors limits the process.

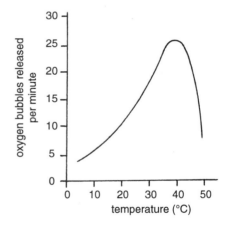

Figure 11.6: Effect of temperature on photosynthetic rate

The graph in figure 11.6 shows that photosynthetic rate rises to an optimum at around 35°C (though this varies from species to species). Beyond the optimum, the photosynthetic rate drops rapidly. This is because photosynthesis consists of many reactions controlled by enzymes which are denatured at high temperatures.

KEY QUESTIONS

1 State THREE ways in which rate of photosynthesis can be measured.

2 **a)** State TWO environmental factors which can be demonstrated (by the bubbler experiment) to affect the rate of photosynthesis by the water-weed *Elodea*.
b) As the distance between the powerful lamp and the *Elodea* plant is gradually decreased, what effect does this have on:
(i) the intensity of the light reaching the plant?
(ii) the number of bubbles of oxygen released by the plant per minute?

3 Consider the information in table 11.2.

ingredients needed to make one loaf	Baker A's stock	Baker B's stock
500 g flour	5 kg flour	5 kg flour
30 g fat	60 g fat	300 g fat
10 g yeast	50 g yeast	50 g yeast
5 g sugar	40 g sugar	15 g sugar

Table 11.2   Factors limiting bread making

a) Which ingredient limits Baker A's bread production to two loaves?
b) Which ingredient limits Baker B's bread production to three loaves?

4 **a)** What is meant by the term *limiting factor*?
b) Identify TWO factors that can limit the process of photosynthesis.

5 When no other factors are limiting the process of photosynthesis, what effect on photosynthetic rate is brought about by raising the temperature of the plant from (i) 15°C to 35°C (ii) 35°C to 55°C?

SPECTRUM OF WHITE LIGHT

When a beam of white light is passed through a glass prism, the **spectrum** of white (visible) light is produced. White light is found to consist of several colours each with a different wavelength (see figure 11.7).

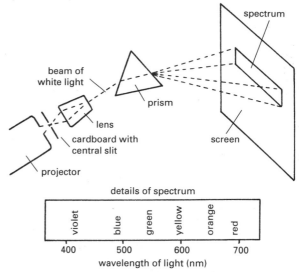

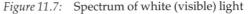

Figure 11.7:   Spectrum of white (visible) light

INVESTIGATING THE EFFECT OF VARYING WAVELENGTH

Figure 11.8 shows an alternative method of measuring *Elodea's* photosynthetic rate. Bubbles of oxygen are allowed to collect in the bend of the tube at point X for a known length of time (e.g. 5 min.). The syringe is then used to draw the gas into the straight part of the tube where it is measured. It is then drawn into the reservoir to allow room for the next bubble and so on.

During this experiment, the wavelength of light is varied using coloured filters and the plant is allowed to become acclimatised to each filter before results are taken. An excess of carbon dioxide is supplied to the plant, a constant high light intensity is maintained and a large water bath is used to surround the test tube. These precautions ensure that wavelength of light is the only **variable factor**. A typical set of results is shown in table 11.3. From these results it is concluded that blue light is the most useful to the plant for photosynthesis.

colour of light allowed through by filter	wavelength of light (nm)	volume of gas released in 5 min (arbitrary units)
blue	430	15.7
green	550	1.2
red	640	8.4

Table 11.3   Typical results for effect of varying wavelength

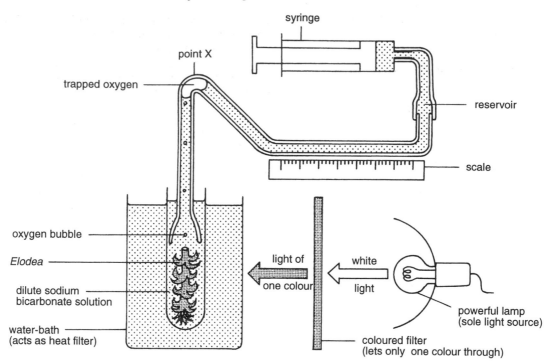

Figure 11.8: Investigating the effect of varying wavelength of light

KEY QUESTIONS

1 **a)** By what means can white light be split up into the spectrum of white (visible) light?
b) Which of the colours in the spectrum consists of light with the longest wavelength?

2 **a)** By what means is photosynthesis measured in the experiment shown in figure 11.8?
b) What is the one variable factor being investigated in this experiment?
c) Name THREE other environmental factors that were kept constant in this experiment and state how this was done in each case.
d) Which colour of light was (i) best (ii) poorest at promoting photosynthesis?

RESPIRATION

Respiration is the name given to the release of **energy** from food in living cells. Respiration occurs in all living cells, twenty-four hours a day.
The process is summarised by the following word equation:

$$\text{glucose} + \text{oxygen} \xrightarrow[\substack{\text{controlled} \\ \text{stages}}]{\text{many enzyme-}} \text{carbon dioxide} + \text{water} + \text{energy}\ (\text{e.g. heat})$$

IMPORTANCE OF RESPIRATION

Respiration is essential to all living cells because it is the means by which energy is made available for:

- cellular work (e.g. active uptake of mineral salts from the soil by root hairs),
- the building up of complex molecules (e.g. proteins and nucleic acids) from simpler molecules,
- growth and reproduction.

KEY QUESTIONS

1 What is meant by the term *respiration*?

2 In which type of plant cells does respiration occur?

3 Summarise the process of respiration as a simple word equation.

4 Give TWO reasons why respiration is of vital importance to plant cells.

MEASURING RATE OF RESPIRATION

Respiratory rate can be estimated by measuring one of the following:

- **oxygen consumption** per unit time,
- **carbon dioxide production** per unit time,
- **energy production** per unit time.

The first of these can be measured using a **respirometer** (see figure 11.9). Carbon dioxide given out by the respiring bean seeds (of known weight) is absorbed by the sodium hydroxide. Oxygen taken in by the respiring bean seeds causes a **decrease** in volume of the enclosed gas. As a result, level B rises in the manometer.

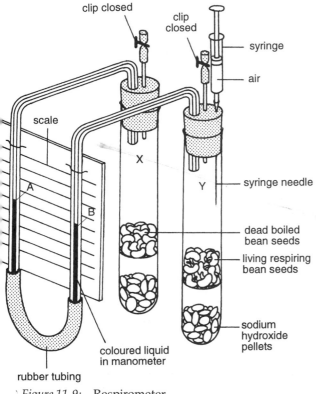

Figure 11.9:   Respirometer

After a known length of time, the syringe is used to find out the volume of air which must be

injected to return the coloured liquid to its initial level. This is equal to the volume of oxygen consumed by the bean seeds.

Rate of respiration can then be expressed as volume of oxygen used per gram of fresh tissue per unit time.

FACTORS AFFECTING RATE OF RESPIRATION

Several environmental factors affect the respiratory rate of a plant tissue. These include oxygen concentration, pH and temperature.

INVESTIGATING THE EFFECT OF VARYING TEMPERATURE

The respirometer is set up as shown in figure 11.9 with the clips left open at the start. Once the live and dead bean seeds have been placed in their tubes, the respirometer is hung onto the side of a thermostatically controlled water bath so that tubes X and Y are suspended in the water (at the required temperature) and the manometer and scale are outside the bath.

The apparatus is allowed to equilibrate and then, with the two manometer levels A and B equal, the clips are closed simultaneously. Readings are taken over a period of time and rate of oxygen consumption at this temperature calculated. The experiment is repeated at different temperatures.

Figure 11.10 shows a graph of a typical set of results. The optimum temperature is found to be

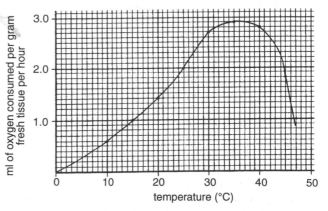

Figure 11.10:   Effect of temperature on respiration rate

around 35°C. This value is not exactly the same for all higher plant tissues but the optimum is normally found to be between 30° and 40°C. The rate of respiration decreases rapidly at higher temperatures as enzymes become denatured.

Respiration versus photosynthesis

If a plant's daily rate of respiration is greater than its daily rate of photosynthesis, it can often survive by falling back on its stored food reserves. However if this state of affairs continues for a long period, the plant will eventually starve to death.

In a natural ecosystem such as grassland, the rate of photosynthesis normally does not change very much over a temperature range of 15°C to 25°C because the low concentration of carbon dioxide in atmospheric air acts as a limiting factor.

However the rate of respiration tends to **double** for an increase in temperature from 15°C to 25°C. The net daily gain in photosynthetic products is therefore reduced at the higher temperatures if carbon dioxide is limiting photosynthesis.

Studies of the potato plant show 20°C to be the optimum temperature for tuber formation. At higher temperatures too much of the photosynthetic product is consumed in respiration. Thus potato plants are cultivated most successfully in cooler climates.

KEY QUESTIONS

1 a) State THREE sets of measurements that can be taken to calculate rate of respiration.
b) Which of these is measured using the respirometer shown in figure 11.9?
c) (i) Does the coloured liquid in the manometer move up towards the living organism or down and away from it? (ii) Explain your answer.
d) Briefly describe the role of the syringe in the respirometer experiment.

2 What apparatus in addition to that shown in figure 11.9 is required to investigate the effect of temperature on rate of respiration?

3 a) Over what range of temperature does respiration normally occur at an optimum rate in higher plant tissues?
b) Why does respiratory rate decrease to zero at higher temperatures?

TRANSPIRATION

Transpiration is the process by which water is lost by evaporation from the aerial parts of a plant. Most transpiration occurs through holes in the leaves called **stomata** (singular = stoma).

STOMATA

Each stoma is surrounded by two **guard** cells. Guard cells differ from normal epidermal cells in three ways. Guard cells are sausage-shaped and have chloroplasts. In addition the inner regions of their cell walls (i.e. those facing the stomatal pore) are thicker and less elastic than the outer regions of their cell walls (see figure 11.11).

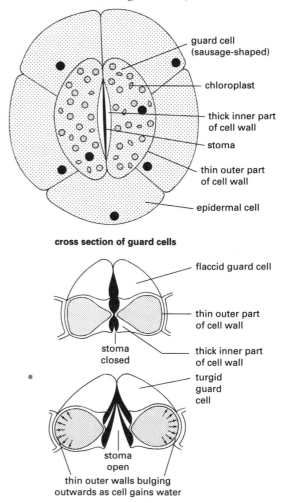

Figure 11.11: Stomata and stomatal mechanism

Stomatal movement

Opening and closing of stomata occur as a result of changes in **turgor** of guard cells.

When water enters a pair of flaccid guard cells, turgor increases. Due to their larger surface area and greater elasticity, the thin outer parts of the two guard cells' walls become stretched more than the thick inner parts. As the two guard cells bulge out, the thick inner walls become pulled apart **opening** the stoma. This happens in light.

When water leaves a pair of guard cells, they lose turgor and return to a flaccid condition. This results in the **closing** of the stoma and takes place in darkness.

ADVANTAGES OF TRANSPIRATION

When stomata are open, water is lost by transpiration. This causes the water concentration of the leaf cells nearest the stomata to drop. In order to replace losses, these cells draw water from neighbouring cells which in turn draw water from the xylem vessels.

This sets up a **transpiration pull** which is the major force drawing water up a plant from the roots. Without transpiration exerting this pull, water and mineral salts would not rise more than a few metres up a plant. Green leaves above this height would fail to receive water for photosynthesis and maintenance of turgor. Tissues in the upper parts of the plant would fail to get supplies of essential minerals for growth.

A further advantage of transpiration is that it keeps the leaves **cool** in hot weather. This works in a similar way to sweating in humans. Excess heat is used to change liquid water into water vapour giving a cooling effect.

DISADVANTAGES OF TRANSPIRATION

Excessive transpiration during a dry sunny spell creates a potential problem: the amount of water lost at the leaf surfaces may be greater than the amount absorbed by the roots.

Under such circumstances leaf cells lose turgor and the stomata close, helping to conserve water. However in many plants a little water continues to be lost **directly** through the outer cells of the leaf. If the dry spell continues the plant starts to **wilt** and may eventually die.

KEY QUESTIONS

1 Define the term *transpiration*.

2 **a)** What is a stoma?
   b) Draw up a table to show THREE differences between a guard cell and a neighbouring epidermal cell.

3 **a)** Does a stoma open or close at nightfall?
   b) With reference to the turgor of guard cells, give a simple description of stomatal movement at daybreak.

4 State TWO ways in which transpiration may be of advantage to a plant.

COMPARING TRANSPIRATION RATES USING A POTOMETER

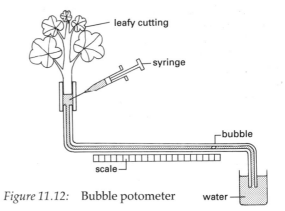

Figure 11.12: Bubble potometer

A bubble **potometer** (see figure 11.12) is an instrument used to measure **rate** of water uptake by a leafy shoot. This rate of water uptake is only approximately equal to transpiration rate since some water may be retained by the leafy shoot for other processes (e.g. photosynthesis).

The open end of the capillary tube is lifted from the beaker of water to allow an air bubble to enter. Once the bubble has appeared on the horizontal arm of the potometer, its rate of movement along the scale is measured (e.g. in mm/min.).

The syringe is used to inject water and return the bubble to the start of the scale allowing the experiment to be repeated for each environmental condition. Table 11.4 gives a specimen set of results for three different environmental conditions.

environmental condition	additional apparatus needed to create condition	average rate of movement of bubble (mm/min)
normal day	none	5
windy day	electric fan	20
humid day	transparent plastic bag	1

Table 11.4   Bubble potometer results

SUMMARY OF FACTORS AFFECTING TRANSPIRATION

Wind

From the above experiment it is concluded that **wind** increases rate of transpiration. This occurs because the air outside the stomata is continuously being replaced with drier air which accepts more water vapour from the plant.

Humidity (amount of water vapour in air)

Increased **humidity** of the air surrounding the plant results in decreased transpiration rate. This occurs because the concentration gradient of water vapour between the inside and the outside of the leaf is decreased. Rate of diffusion of water molecules therefore slows down.

Temperature

Transpiration rate increases with increase in **temperature** due to the faster evaporation rate of water molecules.

Soil water

When **soil water** is in short supply, water concentration of leaf cells including guard cells decreases and they lose turgor. The stomata therefore close and transpiration rate drops to a low level.

Light

Stomata open in **light** and transpiration increases. The reverse occurs in darkness.

KEY QUESTIONS

1 **a)** (i) What does a bubble potometer measure?
(ii) Why is this measurement only approximately equal to transpiration rate?

b) Briefly describe the function of the syringe in the potometer in figure 11.12.

2 Explain why each of the following features of good scientific practice should be adopted when using a bubble potometer:
a) time allowed for plant to get used to each change in environmental conditions before measurements are taken;
b) all factors kept equal except for one change in environmental conditions.

3 List FOUR environmental factors which affect transpiration rate and, for each one, state whether a decrease in the factor leads to an increase or a decrease in rate of transpiration.

TRANSLOCATION

During photosynthesis, a green plant produces carbohydrate food. Some of this is stored as large insoluble molecules of starch; some is transported as smaller soluble molecules of sugar to other parts of the plant (e.g. roots and growing points). This transport of soluble carbohydrates in a plant is called **translocation.**

Higher plants have two different transport systems which are present in vascular bundles in their stems (see figure 11.13). Xylem is located to the inside of each bundle and is found to consist of hollow dead tubes. It is the site of water transport in a plant. **Phloem** is located to the outside of each bundle and consists of living sieve tubes and companion cells.

KEY QUESTIONS

1 **a)** What name is given to the transport of soluble carbohydrates in a plant?
b) Give an example of a soluble carbohydrate that is transported through a plant.
2 **a)** Name the TWO transport tissues found in a higher plant.
b) (i) Which of these is composed of two types of living cells? (ii) Identify ONE way in which each of these 'cells' differs from a typical non-green mature plant cell (e.g. onion epidermal cell).

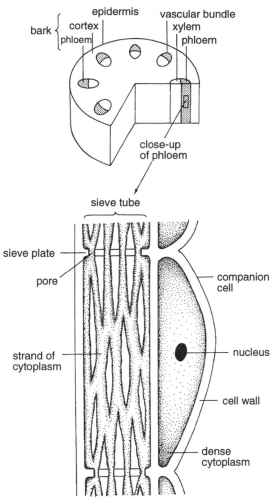

Figure 11.13: Site of phloem in a stem

IDENTIFYING THE SITE OF TRANSLOCATION

RINGING EXPERIMENTS

The experiment shown in figure 11.14 demonstrates that the removal of a complete **ring** of bark (containing the **phloem** tissue) prevents sugar transport in the plant. It is therefore concluded that phloem (or some other tissue in the bark) is the site of sugar transport (translocation).

APHIDS

An aphid (greenfly) is an insect with a long sucking proboscis like a syringe needle which it uses to pierce plant tissues and suck out juices.

Figure 11.15 shows the results of removing an aphid but leaving its proboscis intact. The proboscis is found to be in contact with the **phloem** tissue and to give out sugary liquid. This shows that the phloem is the site of translocation.

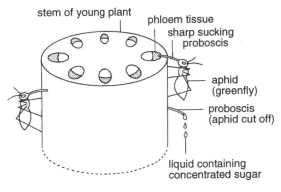

Figure 11.15: Aphids

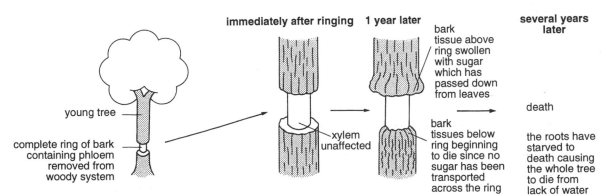

Figure 11.14: Ringing experiment

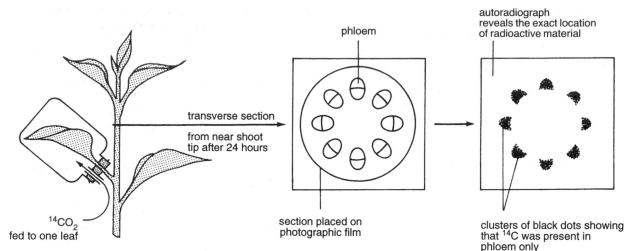

Figure 11.16:   Use of ¹⁴C (in CO₂) as a tracer

RADIOACTIVE TRACERS

Figure 11.16 shows one leaf of a young plant being exposed to carbon dioxide containing an unusual form of **carbon** (**¹⁴C**) which is **radioactive.** This ¹⁴CO₂ is used by the plant during photosynthesis and the ¹⁴C becomes incorporated into sugar molecules.

Twenty-four hours later when a transverse section of the stem is placed in contact with photographic film, the radioactive ¹⁴C atoms (in sugar) give out radiation and cause clusters of black dots to appear on the film. The result is called an **autoradiograph.** The locations of the black marks on it are found to coincide exactly with those of the **phloem** tissue showing that phloem is the site of sugar transport.

Direction of translocation

Figure 11.17 shows the outcome of injecting one leaf of a young plant with sugar containing radioactive ¹⁴C atoms.

The autoradiograph formed twenty-four hours later shows that the 'labelled' sugar has been translocated **up** and **down** the plant. In addition most of the sugar has been transported from the older leaf that was injected to **young leaves** and **fruit**, but not to other older leaves.

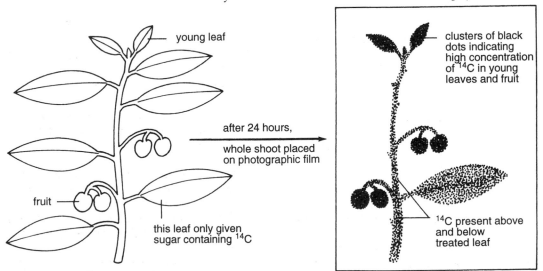

Figure 11.17:   Use of ¹⁴C (in sugar) as a tracer

Mechanism of translocation

Scientists remain uncertain about how translocation works. Its mechanism does seem to be active and to require **energy**. Evidence to support this comes from experiments which show that phloem tissue exhibits a high rate of respiration and that any factor which affects phloem's **metabolic activity** also affects translocation. Low temperatures, respiratory poisons and shortage of oxygen are all factors which reduce rate of translocation.

KEY QUESTIONS

1 a) What group of tissues is removed from a tree during the process of ringing?
b) Which transport tissue does the ring contain?
c) A year after ringing, the tissues above the ring are found to differ from those below the ring. Describe TWO of the differences.
d) Why does a ringed tree normally die after a few years?

2 a) What is an aphid and how does it feed?
b) Describe how aphids can be used to identify the site of sugar transport in a plant.

3 a) Compared with a normal atom of the element carbon, what is unusual about a ^{14}C atom?
b) (i) If a plant is exposed to $^{14}CO_2$ during photosynthesis, which photosynthetic product is later found to contain ^{14}C?
(ii) How can this be used to identify the site of translocation?

4 a) In what direction(s) does translocation occur in a plant?
b) Name TWO environmental factors that can affect rate of translocation.

EXERCISES

1 Match the terms in list **X** with their descriptions in list **Y**.

list X	list Y
1) carbo-hydrate	**a)** dead tissue which transports water up a plant
2) carbon dioxide	**b)** general term for a factor whose limited supply prevents an increase in rate of a process
3) chlorophyll	**c)** apparatus used to measure an organism's rate of respiration
4) limiting factor	**d)** living tissue which transports sugars up and down a plant
5) oxygen	**e)** gas taken in for use during photosynthesis and given out as a product of respiration
6) phloem	**f)** loss of water vapour from a plant's aerial surfaces by evaporation
7) photo-synthesis	**g)** apparatus used to measure a plant's rate of water uptake
8) potometer	**h)** high energy foodstuff produced by photosynthesis
9) respiration	**i)** green pigment present in plants which absorbs light energy
10) respiro-meter	**j)** production of food and oxygen by green plants
11) trans-location	**k)** gas taken in for use during respiration and given out as a product of photosynthesis
12) trans-piration	**l)** release of energy from a food
13) xylem	**m)** transport of soluble carbohydrates in a plant

2 Which of the following word equations correctly represents the process of photosynthesis?
A water + carbon dioxide $\longrightarrow$ sugar + oxygen + energy
B carbon dioxide + water + energy $\longrightarrow$ sugar + oxygen
C sugar + carbon dioxide $\longrightarrow$ water + oxygen + energy
D carbon dioxide + sugar + energy $\longrightarrow$ water + oxygen
(Choose ONE correct answer only.)

3 The graph in figure 11.18 shows the variation in carbon dioxide concentration of the air amongst the leaves of a potato crop during three days in summer.
a) On which day and between which times (to the nearest hour) did the concentration of CO_2 drop at the fastest rate?
b) Which physiological process was responsible for each decrease in CO_2 concentration in the graph?
c) At approximately what time did (i) day break (ii) night fall?

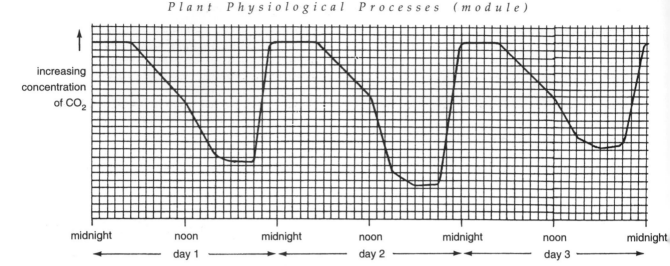

Figure 11.18

d) Suggest which day was probably the least sunny. Explain your answer.

e) Redraw the axes and then make a simple sketch to give an idea of how the CO_2 concentration would be likely to vary during three dull cold days in winter.

4 Table 11.5 refers to the uptake of carbon dioxide by two green plants.

plant	total volume of CO_2 entering plant daily (mm³)	number of leaves	average area of one leaf (mm²)
X	24000	12	500
Y	40000	5	1000

Table 11.5

a) Calculate the daily rate of diffusion of CO_2 into plant X (in mm³ CO_2/mm² of leaf).

b) By how many times is the daily rate of diffusion of CO_2 into plant Y greater than that into plant X?

5 The plant in figure 11.19 was destarched before being set up as shown. This demonstration really consists of three separate experiments being done on the plant at the same time.

After the plant has been in bright light for three days, which two discs should be tested for starch in order to show that:

(i) light is essential for photosynthesis?

(ii) chlorophyll is essential for photosynthesis?

(iii) carbon dioxide is essential for photosynthesis?

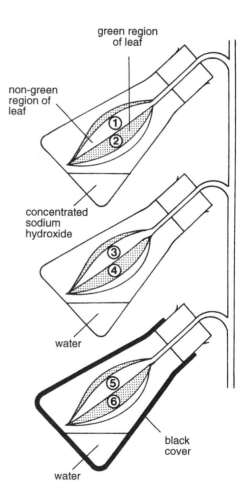

Figure 11.19

6 Table 11.6 refers to the results from an experiment set up to investigate the effect of different wavelengths of light on the rate of photosynthesis in *Elodea*.

colour of light	wavelength (nm)	bubbles of oxygen released/min
blue-violet	400	34
blue	430	52
blue-green	480	35
green	550	10
yellow	590	20
red-orange	640	28
red	690	1

Table 11.6

a) Present the data as a line graph (in the form of a curve).
b) Draw TWO conclusions from the results.
c) Suggest how the experiment could be improved to give a more accurate indication of the optimum wavelength of light for photosynthesis.

7 The graph in figure 11.20 presents the results from two experiments on the rate of photosynthesis by a water plant exposed to different environmental conditions. The CO_2 concentration of pond water was maintained at a high optimum level throughout the two experiments.

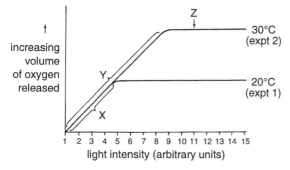

Figure 11.20

a) In experiment 1, the temperature was kept constant at 20°C.
(i) Name the environmental factor that was varied.
(ii) Name a second environmental factor that was kept constant.
b) In the second experiment, the temperature was kept constant at 30°C. Suggest how this was done.

c) What factor was limiting the rate of photosynthesis at region X on the graph?
d) State the light intensity at which a temperature of 20°C first began to limit photosynthetic rate.
e) State the factor which limits photosynthetic rate at (i) region Y (ii) point Z on the graph for experiment 2.

8 The apparatus shown in figure 11.21 was set up in an attempt to show that wheat grains produce carbon dioxide during respiration.

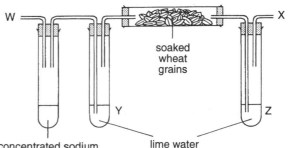

Figure 11.21

a) Spot the error in the experimental set-up and state how you would correct it.
b) Assume that the error has been corrected. Should the pump draw air through the apparatus from W to X or from X to W?
c) If the wheat grains are giving out carbon dioxide, in which flask will the lime water turn milky? Explain your answer.

9 The graph in figure 11.22 shows the rate of uptake of oxygen by a plant in darkness at different temperatures.
a) (i) At what temperature was a volume of 60cm³ of oxygen per day being taken up?
(ii) Identify the physiological process

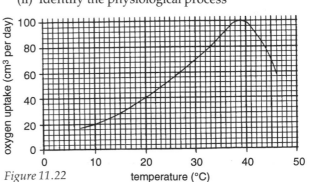

Figure 11.22

responsible for uptake of oxygen by the plant.
b) What volume of oxygen was taken up at 15°C?
c) At what temperature was the highest volume of oxygen taken up?
d) By how many times was the volume of oxygen taken up at 30°C greater than that taken up at 10°C?
e) What increase in temperature was required to increase the volume of oxygen taken up at 20°C to double that amount?

10 The graph in figure 11.23 shows the changes in oxygen concentration in the water of two marine rock pools during a sunny summer's afternoon and early evening. Pool X was shaded in the afternoon from 1400 hours onwards. Pool Y was shaded until 1400 hours.

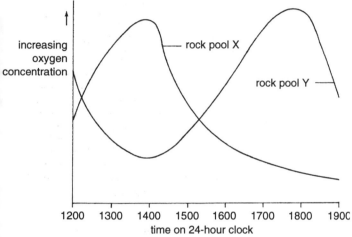

Figure 11.23

a) Identify the dominant physiological process occurring between 1200 and 1400 in (i) pool X (ii) pool Y.
b) Identify the dominant physiological process occurring between 1400 and 1700 in (i) pool X (ii) pool Y.
c) Identify the dominant physiological process occurring between 1800 and 1900 in (i) pool X (ii) pool Y.
d) At which time of day shown on the graph would the water from pool Y be richest in carbon dioxide? Explain your answer.
e) In what way would the graph for pool X have differed if the day had been cloudy between 1200 and 1400 hours?

11 The apparatus shown in figure 11.24 was set up to measure the respiratory rate of dandelion leaf tissue at two different temperatures. Metal foil was placed round both tubes and they were maintained at 15°C for two hours. The experiment was repeated at 25°C. Table 11.7 summarises the results.

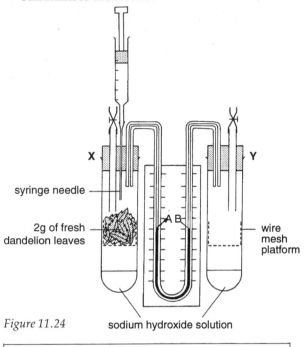

Figure 11.24

temperature (°C)	ml air needed after 2 hours to return level A to starting point
15	2.8
25	4.8

Table 11.7

a) Describe how tubes X and Y could be kept at the required temperature during each part of the experiment.
b) Why were tubes X and Y covered with metal foil?
c) Why does level A rise during the experiment?
d) By what means is level A returned to its starting point after two hours in each experiment?
e) Calculate the respiratory rate (in ml oxygen consumed per gram of fresh tissue per hour) of dandelion leaves at the two temperatures.
f) What conclusion can be drawn from the data about the effect of temperature on respiratory rate of dandelion leaf tissue?

12 The graph in figure 11.25 shows the loss of water from a leafy green plant over a 24-hour period in spring.

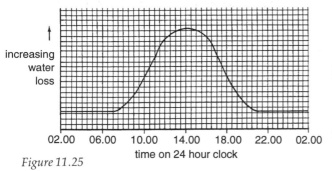

Figure 11.25

a) Name the physiological process that is responsible for the loss of water in the graph.
b) Suggest why the shape of the graph changes between the following times:
(i) 0700 and 1400
(ii) 1400 and 2100.
c) At which time of day did the stomata begin to open? Suggest why.
d) (i) Between which TWO times given in part b) were the stomata most likely to be completely closed?
(ii) Explain why.

13 The bar graph in figure 11.26 shows the transpiration rates of three species of *Pelargonium* (W, X and Y) at air temperatures 15°C, 20°C and 25°C.

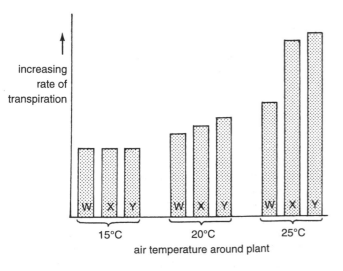

Figure 11.26

a) What relationship exists between transpiration rate and air temperature for all three species of *Pelargonium*?
b) (i) Assuming that all three species have the same rate of water intake, which one is least likely to wilt during a spell of hot dry weather?
(ii) Explain your answer.

14 Horse chestnut trees often grow to a height of 25 metres above ground level. In what way is a rapid rate of transpiration of importance to the continued survival of these plants?

15 The graph in figure 11.27 shows the relationship between the external air temperature and average leaf temperature of two different plant species (A and B) growing beside one another in a greenhouse.

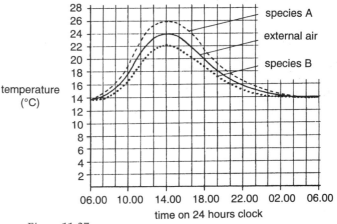

Figure 11.27

a) At what time of day was the temperature of the air in the greenhouse at its maximum?
b) (i) Which species of plant was able to keep cool by transpiring?
(ii) With reference to the graph, explain how you arrived at your answer.
c) (i) Predict which plant would be first to suffer wilting if soil water were in short supply during the day.
(ii) Explain your answer.

16 The graph in figure 11.28 shows the rates of water absorption and transpiration that occurred in a sunflower plant during a 24-hour period.

From this evidence it is true to say that
A in light, absorption rate always exceeds transpiration rate.

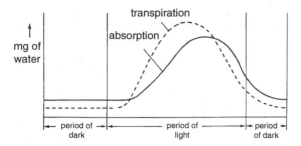

Figure 11.28

    B in dark, absorption rate always exceeds transpiration rate.
    C in light, transpiration rate always exceeds absorption rate.
    D in dark, transpiration rate always exceeds absorption rate.
    (Choose ONE correct answer only.)

17 The data in table 11.8 refer to a series of experiments in which a large leafy shoot attached to a bubble potometer was subjected to various conditions of several abiotic factors.
    a) Consider each of the following pairs of experiments in turn and then draw a conclusion about the effect of a named abiotic factor on the time taken by the bubble to travel 100 mm:
    (i) 1 and 5
    (ii) 3 and 4
    (iii) 6 and 10
    (iv) 5 and 7.
    b) Why is it impossible to draw a valid conclusion about the effect of an abiotic factor on time taken by the bubble from a comparison of experiments 3 and 5?
    c) Which experiment in the table should be compared with experiment 5 to find out about

the effect of darkness on time taken by the bubble to travel 100 mm?
    d) Which TWO experiments should be compared in order to find out the effect of wind speed on time taken by the bubble when the plant is in light at 25°C and in air of 95% humidity?
    e) Which TWO experiments should be compared in order to find out the effect of temperature on transpiration rate by the plant when exposed to wind speed of 15 m/s in light and in air of 75% humidity?

18 The plant shown in figure 11.29 was kept in bright light for two days and then ringed.

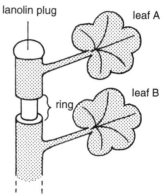

Figure 11.29

    a) (i) What vascular tissue is undamaged by ringing?
    (ii) What substance does it transport?
    b) (i) What vascular tissue is removed by ringing?
    (ii) What substance does it normally transport?
    c) After ringing, the plant was left in darkness for 24 hours and then tested for starch. Leaf A gave a positive result, leaf B a negative one. Explain why.

		experiment number									
		1	**2**	**3**	**4**	**5**	**6**	**7**	**8**	**9**	**10**
abiotic factor	wind speed (m/s)	0	0	0	0	15	15	15	15	15	15
	temperature (°C)	5	25	5	25	5	25	5	25	5	25
	air humidity (%)	75	75	95	95	75	75	95	95	75	75
	light (L)/dark (D)	L	L	L	L	L	L	L	L	D	D
time taken by bubble to travel 100 mm		3 min 3 s	1 min 35 s	4 min 28 s	3 min 2 s	1 min 56 s	0 min 30 s	3 min 22 s	1 min 57 s	24 min 10 s	22 min 4 s

Table 11.8

19 The plant shown in figure 11.30 was ringed one week prior to the aphid invasion. Assuming that no leaves are present below the ring, which aphid will obtain the largest supply of sugar?

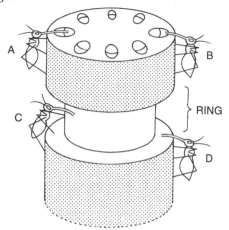

Figure 11.30

20 The entire plant shoot shown in figure 11.31 was exposed to carbon dioxide containing radioactive carbon and this became incorporated into the sugar formed during photosynthesis.

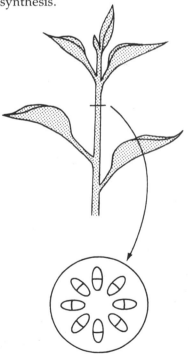

Figure 11.31   T.S. cut after 24 hours

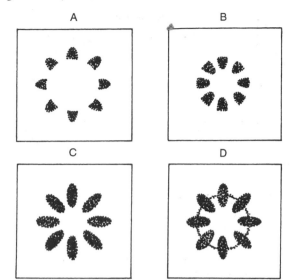

Figure 11.32

Which ONE of the boxes shown in figure 11.32 best represents the autoradiograph that would result from a transverse section of the plant's stem being exposed to photographic film?

21 Early American pioneers frequently ringed trees two or three years before clearing the land for cultivation. Briefly explain the reason for this procedure.

22 The graph in figure 11.33 shows the results of measuring sugar content of leaf cells and stem phloem over a period of 24-hours.
a) At what time of day is the sugar content of the leaves at its:
(i) highest level?
(ii) lowest level?
(iii) Relate this variation in sugar content to a named environmental factor which affects green plants.
b) At what time of the day is the sugar content of the stem phloem at its:
(i) highest level?
(ii) lowest level?
(iii) Why do both of these times occur a little later than those in question a)?
c) (i) What name is given to the transport of sugar in phloem tissue?
(ii) Would it be correct to say that rate of sugar transport is affected by variation in light intensity?
(iii) Explain your answer.

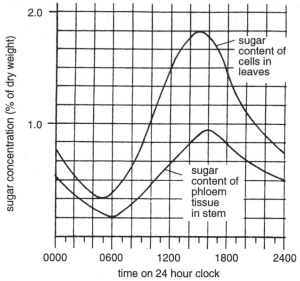

Figure 11.33

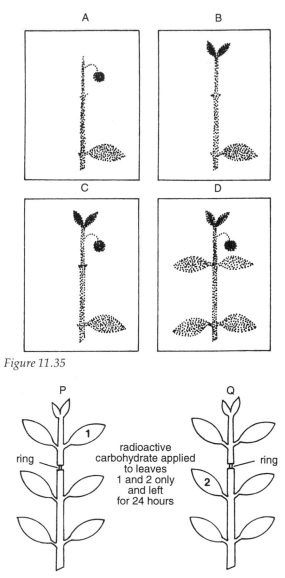

Figure 11.35

23 The plant shown in figure 11.34 was fed carbohydrate containing radioactive carbon at leaf X and left for 24 hours.

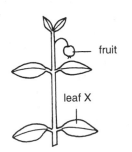

Figure 11.34

Which ONE of the boxes in figure 11.35 best represents the autoradiograph that would result from the plant being exposed to photographic film?

24 Both parts of this question refer to figure 11.36 of two plants, P and Q, and to the four possible autoradiographs A,B,C and D shown in figure 11.37.
a) Which autoradiograph would result on exposing plant P to photographic film?
b) Which autoradiograph would result on exposing plant Q to photographic film?

Figure 11.36

25 Figure 11.38 shows an experiment involving two fruit plants, X and Y. Only the two leaves marked ^{14}C were given radioactive carbon dioxide. The boxed numbers refer to the level of radioactivity (counts/min) found at various parts of the plant several hours later.
a) Name the region to which most radioactive carbohydrate was transported in (i) plant X (ii) plant Y.
b) To which region was the second highest amount of radioactive carbohydrate

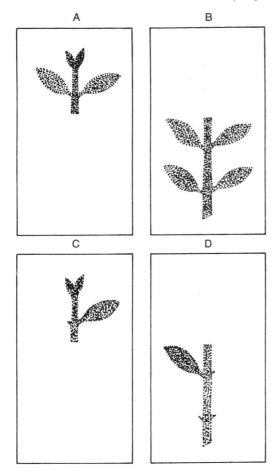

Figure 11.37

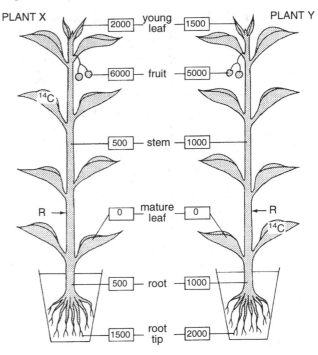

Figure 11.38

transported in (i) plant X (ii) plant Y?
c) What does the experiment demonstrate about the direction in which carbohydrate transport occurs in a plant?
d) Imagine that each plant had first been ringed at point R before these leaves were exposed to $^{14}CO_2$. Make simple diagrams of the autoradiographs that would have resulted if each complete plant had been exposed to photographic film 24 hours after its leaf first received $^{14}CO_2$.

12 RESPONSES TO GROWTH SUBSTANCES

PLANT GROWTH SUBSTANCES

Plant **growth substances** (**hormones**) are chemicals which speed up, inhibit or otherwise affect growth and development of plants.

A hormone is produced at one site in a plant and transported to another site where, in low concentration, it brings about its effect.

AUXINS

Early evidence for the existence of plant growth substances came from work done on the coleoptiles of oat seedlings (see figure 12.1)

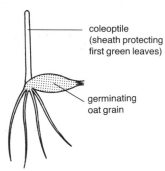

Figure 12.1:   Oat seedling

From the series of experiments shown in figure 12.2 the following conclusions were drawn:
- the shoot tip is essential for growth and produces a chemical;
- this chemical messenger diffuses down to lower regions of the shoot where it stimulates growth by making cells elongate;
- the growth substance is able to diffuse through agar (or gelatin) but not through metal.

The hormone involved is now known to be one of a group of plant growth substances called **auxins.** The most common auxin is **indole acetic acid (IAA).** IAA is produced by root and shoot tips of plants. It is transported away from the tips.

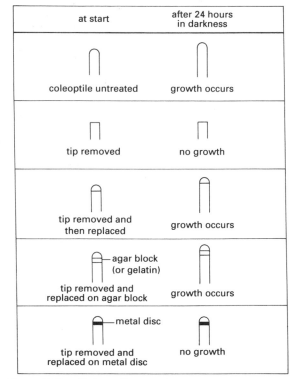

Figure 12.2:   Coleoptile experiments

MECHANISM OF ACTION OF AUXIN IN CELL ELONGATION

Auxin-regulated **genes** have been shown to exist in tissues which undergo cell elongation. These genes respond to the presence of auxin by undergoing an increase in rate of transcription of **mRNA**.

Although a definite link has not yet been established, it is thought that these molecules of mRNA code for **enzymes** which act on the **cellulose fibres** of newly formed cell walls causing them to loosen. Water then enters the cell by osmosis (see figure 12.3) and the cell becomes longer. Since the cell wall becomes stretched irreversibly, this results in the formation of a permanently elongated cell. This process occurs continuously at growing root and shoot tips.

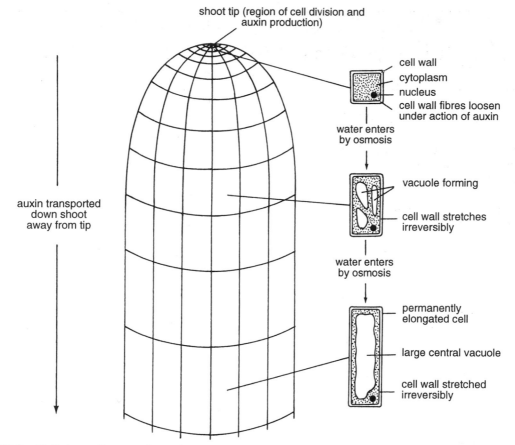

Figure 12.3:   Cell elongation at a shoot tip

KEY QUESTIONS

1 a) What is meant by the term *plant growth substance?*
b) What other term (one word) is also used as a general name for plant growth substances?

2 a) What is an oat coleoptile?
b) Which part of the coleoptile produces a chemical essential for growth and development of the shoot?
c) Which region of the shoot is affected by the chemical?
d) What effect does the chemical have on the cells in this region of the shoot?
e) To which group of plant growth substances does the chemical belong?

3 Construct a flow diagram to summarise the possible mechanism of auxin action in cell elongation.

EFFECTS OF AUXINS AT ORGAN LEVEL
Roots and shoots

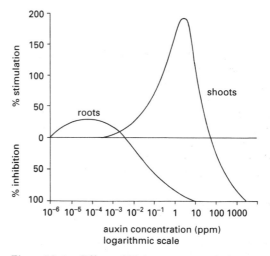

Figure 12.4:   Effect of IAA on roots and shoots

The graph shown in figure 12.4 summarises the results from an experiment where lengths of roots and shoots are treated with various concentrations of auxin. Following treatment the lengths of the organs are compared with those of a control group to assess whether their growth has been **stimulated** or **inhibited**.

It is found that very low concentrations of auxin stimulate elongation of roots but have little or no effect on shoots. On the other hand, higher concentrations of auxin which stimulate shoots inhibit the elongation of roots.

COMMERCIAL APPLICATIONS OF AUXINS
Rooting powders

Roots which arise from a plant part other than the main root (or its branches) are called **adventitious** roots.

When **rooting powder** which contains synthetic auxin is applied to the cut ends of stems, it stimulates the formation of adventitious roots. This makes the plant easy to propagate.

Herbicides

Synthetic auxins are used as **selective weedkillers** (**herbicides**). They work by stimulating the plant's rate of growth and metabolism to such an extent that the plant exhausts its food reserves and dies of starvation.

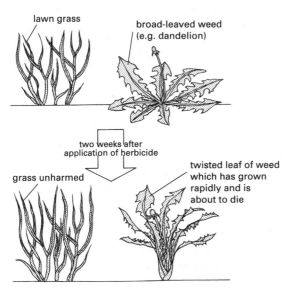

Figure 12.5: Action of selective weedkiller

Selective weedkillers are especially effective on lawns since weeds with broad leaves (see figure 12.5) absorb much of the chemical and die whereas narrow-leaved grass plants absorb little of the chemical and are hardly affected.

Fruit formation

Following fertilisation in a flower, auxin promotes the formation of the fruit coat from the ovary wall. The coat may become soft and succulent as in a grape.

Fruit development **without** fertilisation (called **parthenocarpy**) can be induced artificially by treating unfertilised flowers with auxin. This process produces a plentiful supply of **seedless** fruit that is ripe for harvesting at the same time.

Fruit (and leaf) abscission

Abscission is the separation of fruit (and leaves) from a plant. It is prevented by the presence of a high concentration of auxin in the fruit (or leaf) tissues. Abscission occurs when auxin concentration falls. Fruit growers often spray their crop with synthetic auxin to delay abscission and prevent the fruit from falling too early.

Apical dominance

Very high concentrations of auxin inhibit growth. In many plants, the apical bud is able to inhibit the development of side buds further down the stem by producing a sufficiently high concentration of auxin which is translocated down the phloem tissue to the side buds.

Such auxin-controlled **apical dominance** is demonstrated by the experiment shown in figure 12.6. In the absence of the apical bud (or an auxin-containing substitute), hormone concentration drops to a level which no longer inhibits growth of side buds. These now develop into side branches.

Gardeners often remove apical buds to promote the development of side branches in plants which would otherwise become too tall and spindly.

KEY QUESTIONS

1 Rewrite the following sentences choosing the correct word in each case.
Very low concentrations of auxin stimulate elongation of roots/shoots but have little effect

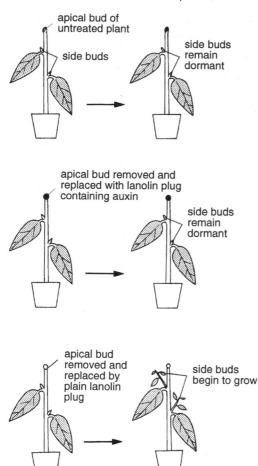

Figure 12.6: Apical dominance

on roots/shoots. Higher concentrations of auxin which stimulate roots/shoots, inhibit elongation of roots/shoots.

2 Most cereal crops are narrow-leaved; the weeds that compete with them tend to be broad-leaved.
a) What type of chemical could be used in an attempt to destroy only the weeds?
b) Explain how this method works.

3 a) What effect does an auxin have on the ovary wall in a flower following fertilisation?
b) By what means can seedless grapes be produced?

4 a) What name is given to the inhibition of side buds by a high concentration of auxin made by the apical bud on the same plant shoot?

b) What happens to the side buds if the apical bud is cut off and replaced with a plug of plain lanolin paste?
c) What happens to the side buds if the apical bud is cut off and replaced with a plug of lanolin paste containing a high concentration of auxin?

GIBBERELLINS

In the 1920s, a Japanese farmer discovered that some of his rice plants, which were infested with a fungus, had grown abnormally tall. This growth effect was found to have been caused by a growth substance (hormone) secreted by the fungus. Since the fungus was called *Gibberella*, the hormone was called **gibberellin.**

Many gibberellins are now known to exist. They make up a second group of plant growth substances which occur naturally in very small amounts in most plants. The most common one is **gibberellic acid (GA).**

Like auxins, gibberellins stimulate **cell division** and **elongation** in stems. In addition gibberellins affect several other aspects of growth and development.

EFFECT OF GIBBERELLIC ACID (GA) ON DWARF PEA SEEDLINGS

Look at the experiment shown in figure 12.7. The dwarf pea seedlings in pot **A** which receive gibberellic acid reach the same height as the normal tall variety in pot **C**. This is due to an increase in length (not number) of their internodes.

The dwarf plants in the control pot **B** show that this effect is not due to the application of lanolin alone.

It is concluded that gibberellin is needed by a plant to make its internodes increase in length. Tall varieties are able to manufacture sufficient gibberellin to make them grow to full height. The dwarf condition (which is inherited) is caused by a shortage of gibberellin due to a genetic deficiency.

Mechanism of action

Microscopic examination of the stem cells of the plants in the experiment in figure 12.7 reveals that gibberellic acid exerts its influence by promoting **cell elongation.** The exact mechanism by which

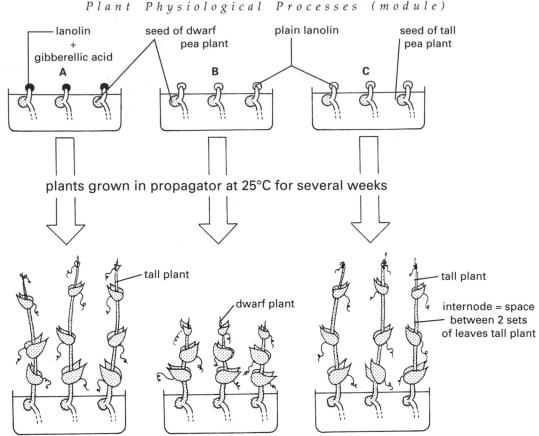

Figure 12.7: Effect of GA on dwarf plants

GA does this remains unclear. In dwarf pea plants, stimulation of growth by GA is accompanied by a large increase in auxin content of the affected stem cells. It is possible that cell elongation is due to enhancement of auxin level brought about by GA.

EFFECT OF GIBBERELLIC ACID (GA) ON GERMINATING BARLEY GRAINS

Figure 12.8 shows the internal structure of a soaked barley grain. It indicates the angle at which the grain should be cut to give an **embryo** 'half' and an **endosperm** 'half. Several of these 'halves' are needed for the experiment shown in figure 12.9.

The results of this experiment show that digestion of starch (in the starch agar) has only occurred in plate **A** beneath the endosperm parts of barley grains.

It is concluded that both endosperm and gibberellic acid must be present for the starch-digesting enzyme, α-**amylase**, to be produced in a barley grain.

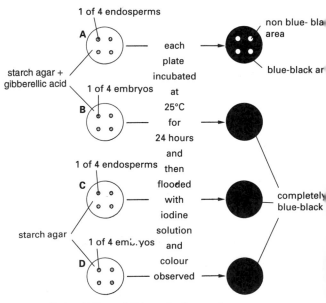

Figure 12.9: Effect of GA on barley endosperm

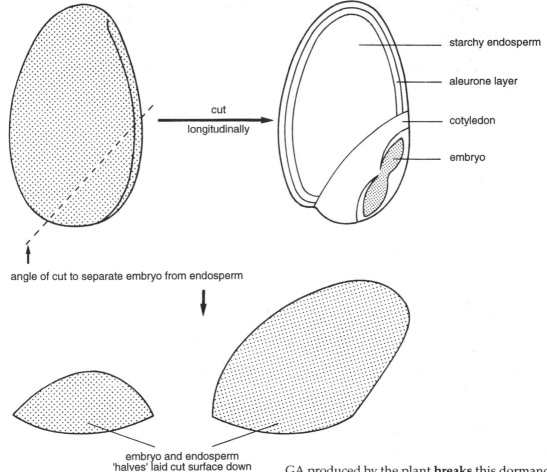

Figure 12.8: Barley embryo and endosperm

Mechanism of action

Under normal circumstances in a soaked cereal grain, the hormone gibberellin is made by the embryo and passes up to the **aleurone** layer (see figure 12.10). Here it acts at **gene** level and **induces** the production of the enzyme α-**amylase.** This digests starch to sugar (maltose) which is needed for growth by the seedling.

By this means, gibberellin breaks the dormancy of many types of seed and induces them to germinate.

EFFECT OF GIBBERELLIC ACID (GA) ON BUD DORMANCY

The buds of deciduous trees remain dormant during winter. Under natural conditions in spring, GA produced by the plant **breaks** this dormancy and allows the buds to open and develop into new leafy branches.

If GA is applied to buds during the winter or early spring, their dormancy is broken artificially and they begin to develop.

KEY QUESTIONS

1 **a)** (i) What effect does gibberellic acid (GA) have on the stem length when applied to the shoots of dwarf pea plants?
(ii) Is this effect due to an increase in number or length of internodes?
b) (i) What does microscopic examination of the cells of the stem internodes reveal about the way GA exerts its influence?
(ii) Why does this effect not occur in the dwarf variety under normal circumstances?

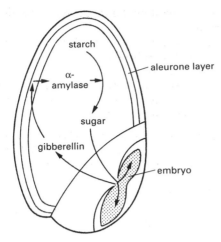

Figure 12.10: Induction of α-amylase by GA

2 a) (i) Name the TWO region of a barley grain separated by the cotyledon. (ii) Which of these contains a store of starch?
b) (i) Which layer of the endosperm can be induced to make a starch-digesting enzyme? (ii) Name this enzyme.
c) (i) What plant growth substance is needed to induce production of the enzyme? (ii) Under normal circumstances, which part of the grain makes the growth substance?
d) (i) Once the enzyme is acting on its substrate, what end product is formed? (ii) Which part of the grain uses this product for growth?

e) Why is it of survival value that the sequence of events only occurs when the grain is soaked and away from extremes of temperature?

3 What effect does GA have on the dormancy of winter buds on deciduous trees?

CYTOKININS

Cytokinins make up a further group of growth substances. They occur in very small quantities in plants. They are most abundant at sites of rapid cell division such as root and shoot tips and growing seeds and fruits.

Cytokinins promote **cell division** and in particular the process of **cytokinesis** (the division of cytoplasm following nuclear division during mitosis). They also play a part in the process of **differentiation** (the alteration and adaptation of unspecialised cells to enable them to perform a particular function as part of a permanent tissue).

SYNERGISTIC INTERACTION

Cytokinins work **synergistically** with auxins. This means that the two growth substances working

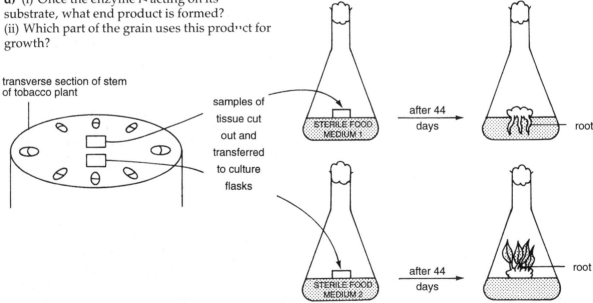

FOOD MEDIUM 1 contains auxin (2 mg/l) and cytokinin (0.02 mg/l)

FOOD MEDIUM 2 contains auxin (0.02 mg/l) and cytokinin (1 mg/l)

Figure 12.11: Tissue cultures

together produce a greater effect than the sum of their individual effects.

Much of what is known about cytokinins comes from work done using tissue cultures. When given certain growth conditions, samples of stem tissue from some plants will undergo cell division forming a mass of undifferentiated cells called a **callus.**

The results of the experiment illustrated in figure 12.11 show that under certain conditions (high ratio of auxin : cytokinin) some of the cells differentiate into root cells whereas under other conditions (low ratio of auxin : cytokinin) some of the cells differentiate into shoot cells.

From these and many other similar experiments, it is concluded that the **genes** that control vital aspects of cell division and differentiation are somehow switched on or off depending on the ratio of auxin : cytokinin present in the cell.

Assuming that these experiments mirror the events that take place naturally in a growing plant, it is clear that cytokinins play a key role in the division and differentiation of its cells.

DELAYING SENESCENCE

During the process of ageing (**senescence**) of a leaf, the levels of DNA and RNA in its cells decrease rapidly as they break down under enzyme action. Cytokinins are found to delay the ageing process. It is thought that they do this by somehow halting the action of the mRNA molecules that code for the enzymes.

BREAKING DORMANCY

Cytokinins also break dormancy in seeds and buds.

KEY QUESTIONS

1 a) Strictly speaking, mitosis means division of the nucleus during cell division. What term refers to the division of cytoplasm following nuclear division?
b) Which group of plant growth substances control this process?
c) Name another aspect of growth at cellular level which is also promoted by this group of plant hormones.

2 a) What is meant by the expression *synergistic relationship*?
b) Name TWO plant growth substances that work together in this way.
c) What is a callus?
d) What balance of the two growth substances is needed for differentiation of callus cells into (i) root cells (ii) shoot cells?

3 a) (i) What happens to the levels of DNA and RNA in the cells of an ageing leaf? (ii) What substances bring about this change in their levels?
b) (i) Which type of plant growth substance delays the ageing process? (ii) By what means is it thought to do this?

4 What effect do cytokinins have on the dormancy of seeds and buds?

ABSCISSIC ACID

Abscissic acid (ABA) is a growth regulator substance which **inhibits** growth of most plant parts. It works by **suppressing** the activity of other growth substances (e.g. auxins). ABA's effect is especially pronounced in certain deciduous trees in autumn. At this time the concentration of ABA in ageing leaves increases to a level high enough to suppress the activity of auxin. The tree becomes prepared for winter by undergoing
- decrease in growth rate of shoots
- development of dormant winter buds
- leaf fall.

LEAF FALL (ABSCISSION)

Abscission is the separation of leaves from a plant. Prior to leaf fall, auxin concentration drops and abscissic acid concentration rises. This hormonal combination leads to the formation of two distinct layers near the base of a leaf stalk as shown in figure 12.12.

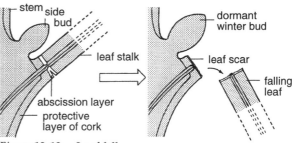

Figure 12.12:   Leaf fall

The **abscission layer** consists of cells whose contents disintegrate and whose walls break down as a result of enzyme action. Eventually the leaf stalk snaps under the weight of the leaf and the leaf falls. The **protective layer** is composed of cork tissue which seals the wound and forms a leaf scar on the stem of the winter twig.

During the winter when the ground may be frozen, it is difficult or even impossible for roots to absorb water. The shedding of leaves in autumn is therefore of survival value to the tree since it greatly reduces the amount of irreplaceable water that would be lost by transpiration.

FRUIT DROP (ABSCISSION)

In some plants, abscission of **fruit** appears to work in a similar way to leaf fall. When the concentration of auxin drops and that of abscissic acid rises, an abscission layer forms in the fruit stalk and the fruit drops.

If a decrease in auxin concentration occurs early in fruit development, this leads to a **premature drop** and loss of much of the fruit crop. This problem can be overcome in many plants by applying auxin sprays to prevent the abscissic acid gaining the upper hand in the interaction between the two opposing plant hormones.

SEED DORMANCY

Abscissic acid **inhibits germination** of some types of seed (see figure 12.13). ABA is thought to suppress the activity of gibberellic acid and therefore prevent the induction of the enzyme α-amylase (see page 142) necessary for germination.

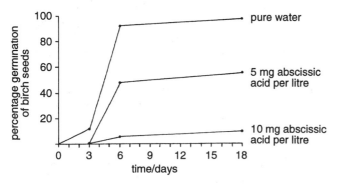

Figure 12.13: Inhibition of birch seed germination

Normally the concentration of ABA drops following a period of cold weather. Winter bud and seed dormancy are broken and growth resumes in the warmer spring weather.

KEY QUESTIONS

1 In what general way is the effect of abscissic acid (ABA) different from that of auxins, gibberellins and cytokinins?

2 State THREE changes that occur in a deciduous tree as winter approaches.

3 **a)** What is meant by the term *abscission?*
   b) What happens to the concentration of
   (i) auxin (ii) ABA in a leaf just before abscission?
   c) Name the TWO layers found at a leaf base at this time and state the function of each.

4 **a)** By what artificial means can fruit fall be delayed?
   b) Why is this delay in fruit fall of advantage to the fruit grower?

5 State the effect of ABA on germination of seeds (e.g. birch tree).

ETHYLENE

Ethylene is a gaseous growth substance produced in minute quantities by several types of plant organ. In many fleshy fruits a rise in ethylene production is found to begin a few days before the onset of ripening as illustrated in figure 12.14 for tomato.

The ethylene is produced by the unripe fruit **itself** and exerts an influence over its own ripening process by enhancing the action of respiratory enzymes. This triggers an **increase** in **respiration rate** which may then remain fairly steady during ripening (e.g. citrus fruits) or rise to a peak and then drop again (e.g. tomato and banana).

The energy generated by the increase in respiratory rate is channelled into the ripening process by being used in the synthesis of **enzymes.** These are required for
  • destruction of chlorophyll and synthesis of red and/or yellow pigments leading to a change in fruit colour
  • breakdown of tough cell walls leading to softening of fruit tissues

- digestion of complex carbohydrates to simple sugars leading to formation of fruit juice with a sweet flavour.

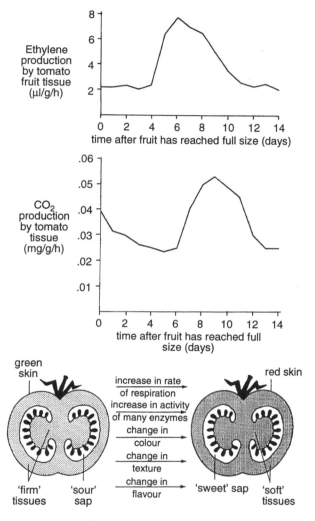

Figure 12.14:   Effect of ethylene on ripening of tomato fruit

STORAGE OF FRUIT

Fully grown but unripe fruit can be successfully stored for long periods if it is prevented from respiring rapidly. This can be done by keeping it in an atmosphere which is low in oxygen and/or rich in carbon dioxide. Under these conditions any naturally synthesised ethylene is unable to bring about its effect and ripening is delayed.

The fruit (e.g. banana) can later be treated with ethylene in normal conditions of high oxygen and low carbon dioxide to promote the ripening process when required for the market.

HORMONE SUMMARY

Table 12.1 summarises the main effects brought about by the five types of growth regulator substance.

hormone	effect
auxins	stimulation of cell division, promotion of cell elongation, inhibition of leaf and fruit abscission, apical dominance
gibberellins	promotion of cell elongation, reversal of genetic dwarfism, breaking of dormancy in seeds and buds
cytokinins	promotion of cell division, delaying of leaf senescence (ageing), breaking of dormancy in seeds and buds
abscissic acid	inhibition of growth, promotion of abscission of leaves and fruit, induction of dormancy in seeds and buds
ethylene	promotion of ripening of fruit

Table 12.1   Summary of effects of plant hormones

KEY QUESTIONS

1 **a)** (i) What effect does ethylene have on the action of respiratory enzymes in unripe fruit? (ii) In what way does rate of cellular respiration change as a result? (iii) What use is made by the cells of the extra energy generated? (iv) State THREE processes that are controlled by the newly synthesised enzymes.

2 **a)** Describe the conditions that are used to prevent fully grown green-picked fruit from becoming ripe.
b) Why do they not ripen under such conditions?
c) What can later be done to the fruit to speed up the ripening process when the fruit is required for the market?

1 Match the plant growth substances in list **X** with their effects given in list **Y**.

list X	list Y
1) abscissic acid	**a)** works synergistically with auxin during cell differentation
2) auxin	**b)** promotes the ripening of fruit
3) cytokinin	**c)** suppresses the activities of other growth substances
4) ethylene	**d)** reverses genetic dwarfism in mutant dwarf varieties
5) gibberellin	**e)** causes apical dominance

2 a) Which hormone induces dormancy in many seeds and buds?
b) Name TWO hormones that can break dormancy of seeds and/or buds.
c) Which TWO hormones promote cell elongation?
d) Name the hormone that (i) promotes (ii) inhibits abscission of leaves.
e) Which hormone makes fruit become ripe?
f) Name the growth substance that delays the ageing (senescence) of leaves.

3 The graph in figure 12.15 summarises the results of an experiment in which the rate of increase in length of an oat coleoptile was followed over a period of ten hours.

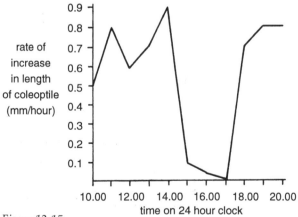

Figure 12.15

a) Name the growth substance made by the tip of an oat coleoptile.
b) (i) What effect does the substance have on the shoot as a whole?

(ii) What effect does the substance have on individual cells?
c) (i) At approximately what time was the coleoptile in the graph decapitated?
(ii) Explain your answer.
d) Suggest why a slight increase in coleoptile length still occurred after decapitation.
e) Name TWO possible courses of action that could have been taken at 1700 hours to bring about the change in rate of increase in length of coleoptile that followed.

4 Table 12.2 shows the results from an investigation into the effect of different concentrations of auxin on roots and shoots of cress seedlings. In this investigation 10mm lengths of root and shoot were immersed in each of the auxin solutions and kept at 30°C.

After 14 hours averages were calculated and the results in table 12.2 obtained.

molar concentration of auxin applied	average length of organ after 24 hours (mm)	
	root	shoot
0	20	20
10^{-5}	22.5	20
10^{-4}	23.5	20
10^{-3}	22	20.5
10^{-2}	18.5	22.5
10^{-1}	14	27.5
1	11	35.0
10^{1}	10	30
10^{2}	10	16

Table 12.2

a) Present the data as two line graphs sharing the same axes.
b) Which concentration of auxin produced the greatest stimulation of (i) roots (ii) shoots?
c) What effect did a molar concentration of 10^{-1} have on the growth of (i) roots (ii) shoots?
d) Which concentration of auxin had the effect of inhibiting growth of both roots and shoots?
e) Which concentration of auxin would be most likely to prevent the eyes (dormant shoot buds) of potatoes from sprouting?

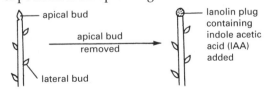

Figure 12.16

5 Both parts of this exercise refer to the experiment shown in figure 12.16 on p.148.

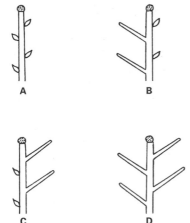

Figure 12.17

a) Which ONE of the diagrams in figure 12.17 shows the appearance of the shoot after two weeks?

b) Which ONE of the diagrams in figure 12.18 shows the most suitable control for this experiment?

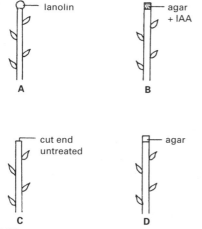

Figure 12.18

6 Figure 12.19 shows the effects of increasing auxin concentration on growth of different plant organs.

a) State the molar concentrations of auxin that cause the maximum growth of each of the four types of plant organ.

b) Which type of plant organ shows least promotion of growth over the whole range of auxin concentration?

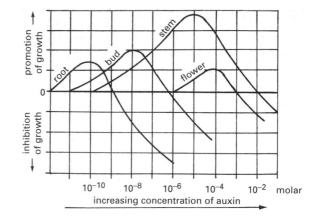

Figure 12.19

c) State TWO molar concentrations of auxin which inhibit growth of roots yet promote growth of both buds and stems.

d) Between which TWO molar concentrations of auxin is stem growth promoted but flower growth inhibited?

7 The graph in figure 12.20 represents the results of an experiment in which four young broad bean shoots were decapitated and then treated as follows:

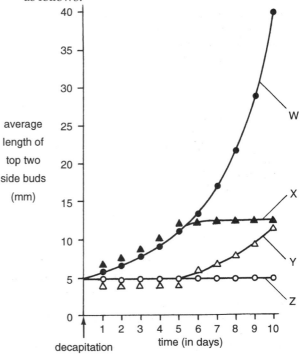

Figure 12.20

(i) auxin applied daily throughout the experiment,
(ii) auxin applied daily until day 5 and then stopped,
(iii) auxin applied daily from day 5 onwards,
(iv) no auxin applied at any time.

During the ten days of the experiment, the lengths of the top two side buds were measured and an average calculated daily.
a) Match (i), (ii), (iii) and (iv) above with W, X, Y and Z in the graph.
b) Justify your choice in each case.

8 Which of the following effects is brought about by BOTH gibberellic acid and indole acetic acid?
A reversal of genetic dwarfism
B breaking of dormancy in seeds
C promotion of cell elongation
D induction of α-amylase in seed grains
(Choose ONE correct answer only.)

9 The bar graph in figure 12.21 shows the results of an investigation into the effect of gibberellic acid on the growth of dwarf pea plants.

The growth substance was applied in lanolin paste to the tip of each shoot 2 days after it had emerged from the germinating seed.

The height of each shoot was measured after 7 and after 22 days of growth.

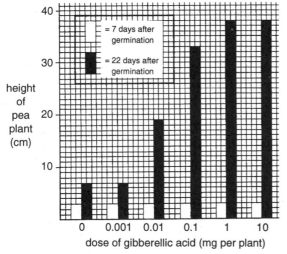

Figure 12.21

a) Make a generalisation about the effect of increasing concentration of gibberellic acid on the height of dwarf pea plants after 22 days of growth.

b) (i) What was the lowest dose of GA needed to give the maximum effect?
(ii) Suggest how you could find out if plants showing this maximum effect are equal in height to the normal tall variety of this strain of pea plant.
c) (i) Which dose of GA had no effect?
(ii) Suggest why.
d) Which treatment was the control?
e) Give a possible explanation for the difference between the results at 7 days and at 22 days.

10 In an experiment, barley embryo and endosperm 'halves' were prepared as shown in figure 12.8 on page 144. These were placed, cut surface down, on plates of starch agar. In some cases gibberellic acid (GA) had also been added to the agar as indicated in table 12.3.

After 24 hours each plate was flooded with iodine solution to locate starch-free areas which indicated digestion. The results are given in the table.

starch agar plate	presence or absence of GA in agar	part(s) of barley grain put on agar	digestion (✓) or no digestion (✗) of starch after 24 hours
1	absent	endosperms	✗
2	absent	embryos	✗
3	absent	endosperms and embryos	✓
4	present	endosperms	✓
5	present	embryos	✗
6	present	endosperms and embryos	✓

Table 12.3

a) What conclusion can be drawn from plates 1 and 4 about the effect of GA on an endosperm's ability to digest starch?
b) What conclusion can be drawn from plates 2 and 5 about the effect of GA on an embryo's ability to digest starch?
c) Compare the contents of plates 3 and 4 and suggest the source of the GA supply in a barley grain under natural conditions of germination.
d) What effect does GA have on endosperm tissue that allows successful digestion of starch to occur?

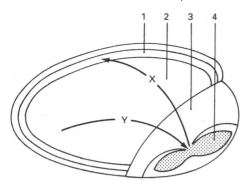

Figure 12.22

e) (i) Which numbered layer of cells in the barley grain shown in figure 12.22 produces the enzyme α-amylase?
(ii) Name this layer.
f) In a germinating barley grain, it is thought that GA travels in direction
A X and digests starchy endosperm to sugar in region 1.
B X and induces α-amylase production in region 1.
C Y and supplies sugar to the cells in region 4.
D Y and induces α-amylase production in region 4.
(Choose ONE correct answer only.)
g) Why is it of survival value to a seed grain that its dormancy is broken by the above mechanism only when water enters the seed grain and the temperature is above freezing point?

11 The graph in figure 12.23 shows the results of an investigation into the effect of three different types of cytokinin on tobacco callus. The initial fresh weight of each callus used was 1g. Each callus was allowed to grow for 40 days.
a) (i) In general what effect did increasing cytokinin concentration have on fresh weight of tobacco callus?
(ii) Suggest how cytokinin brings about this change at cellular level.
b) (i) Which concentration of cytokinin X was most effective at bringing about the response?
(ii) Which type of cytokinin was least effective at the lower concentrations at bringing about the response?
(iii) What is the lowest concentration of cytokinin Y needed to produce a callus with a fresh weight of 18 g?

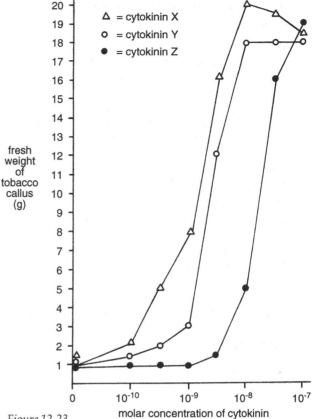

Figure 12.23

c) What additional growth substance is needed to make the undifferentiated callus cells differentiate into roots or shoots?

		concentration of auxin in food medium (mg/l)	
		0	**1.0**
conc. of cytokinin in food medium (mg/l)	0	callus unchanged	callus has increased slightly in size by forming a few large multinucleate cells
	0.02	callus unchanged	callus bears roots
	0.2	callus has increased slightly in size by forming a few uninucleate cells	callus has increased greatly in size by forming many normal-sized uninucleate cells
	1.0	callus has increased slightly in size by forming a few uninucleate cells	callus bears shoots

Table 12.4

12 Table 12.4 summarises the results of an
investigation into the combined effect of two
plant growth substances on the growth of
carrot plant callus after 30 days of growth.
a) Suggest what is meant by the terms
(i) multinucleate (ii) uninucleate.
b) Under what conditions are:
(i) a few large multinucleate cells formed?
(ii) many normal-sized uninucleate formed?
(iii) Suggest the role played by cytokinin as
indicated by these results.
c) What combination of the two growth
substances was required for the callus to
develop (i) roots (ii) shoots?
d) What effect does increasing the
concentration of cytokinin have on the type of
differentiated organ formed when the auxin
concentration is maintained at 1 mg/l?
e) With reference to the table, present support
for the idea that the combined effect of
cytokinin and auxin is greater than the sum of
their separate effects.

13 Figure 12.24 shows the young root growing
from a germinating pea seed. Many such roots
were grown and then cut up into separate
regions 1, 2 and 3.

The contents of the cells in each of these
regions were extracted and applied to barley
seedlings about to show senescence (ageing) of
leaves. Table 12.5 summarises the results.

region of pea root from which cell extract was obtained	effect on barley leaves
1	senescence delayed for long period
2	senescence delayed for short period
3	senescence not delayed

Table 12.5

a) Which plant growth regulator substance
delays the senescence of leaves?
b) (i) Which region of the young pea root is
richest in this growth substance?
(ii) Explain how you arrived at your answer
to (i).
(iii) Suggest the natural role that the substance
normally plays in the region of the root that
you named.
c) Name a second growth substance that plays
a part in the production of the differentiated
side roots at region 3 in the diagram.

14 All the leaf blades were removed from three
healthy plants of the same type leaving the leaf
stalks intact. The cut ends of the leaf stalks
were then treated as shown in figure 12.25.

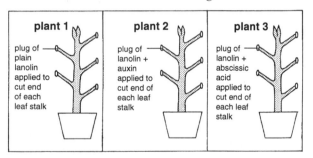

Figure 12.25

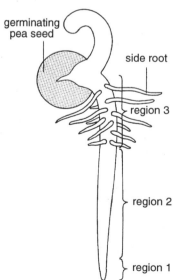

Figure 12.24

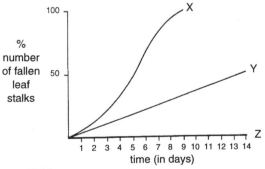

Figure 12.26

The number of leaf stalks that had formed an abscission layer near their bases and had fallen after two weeks was counted for each plant and the results graphed as shown in figure 12.26.

a) Match graphs X, Y and Z with plants 1, 2 and 3.

b) Explain your choice in each case.

15 Table 12.6 refers to concentrations of abscissic acid present in the side buds on a twig of a beech tree.

month	concentration of abscissic acid (g/kg fresh weight)
August	30
September	70
October	100
November	310
December	130
January	110
February	60
March	40
April	20

Table 12.6

a) Present the data as a bar graph.

b) Describe the trend in concentration of abscissic acid that occurs during autumn.

c) What effect does this change in concentration have on the

(i) leaves in autumn?

(ii) state of the buds during November?

d) (i) What has happened to level of abscissic acid by the time spring arrives?

(ii) Predict what will happen to the buds during March and April.

16 Figure 12.27 shows four fruits of the same type which have been treated as indicated. If they had been left untreated they would have been expected to fall naturally in about four weeks' time.

a) (i) Predict which fruit will fall first.

(ii) Give a reason for your choice.

b) (i) Predict which fruit will be the last to fall.

(ii) Give a reason for your choice.

c) Describe the changes in structure that occur at cellular level in a fruit stalk just before fruit fall.

17 Leaves X and Y in figure 12.28 (i) were approaching an age at which they would be lost naturally from each plant by abscission. One of these leaves was then treated with auxin.

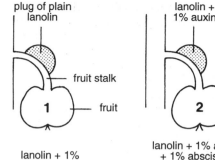
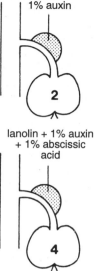

Figure 12.27

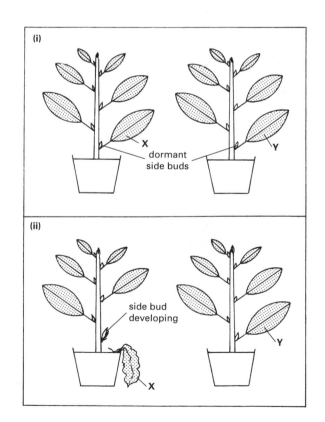

Figure 12.28

Figure 12.28 (ii) shows the appearance of the plants three weeks later.

a) Which leaf was treated with auxin?

b) Give TWO reasons for your answer.

c) Which naturally occurring growth substance could have been extracted from another plant and used to make leaf X fall even sooner that it did?

18 The graphs in figure 12.29 summarise data obtained from studies of a species of fruit tree.

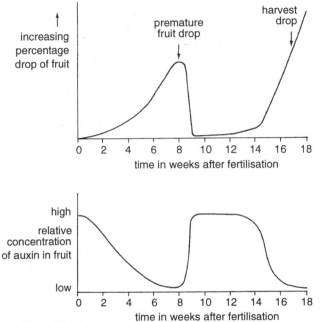

Figure 12.29

a) What general relationship exists between percentage drop of fruit and relative concentration of auxin in fruit over the 18-week period?

b) Between weeks 6 and 12 which week showed the (i) highest (ii) lowest percentage drop of fruit?

(iii) State the relative concentration of auxin present in the fruit at these two times.

(iv) Suggest the relative concentration of abscissic acid that was present in the fruit at these two times.

c) What could the fruit grower spray onto the crop in an attempt to prevent the unwanted premature fruit drop?

d) What could the fruit grower apply to the crop to make all the fruit drop at the same time during harvesting?

19 Figure 12.30 summarises an investigation involving the seeds of a tomato plant.

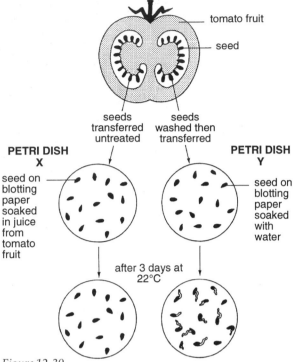

Figure 12.30

a In which dish was seed germination inhibited?

b) (i) Identify the likely source of the inhibitor.

(ii) Name the growth regulator substance that is able to inhibit seed germination in many plants.

c) Why is inhibition of tomato seeds while they are in the fruit of benefit to the plant?

20 Provided that an unripe fleshy fruit such as a tomato, apple or pear has reached its full size, it will still ripen whether left attached to the parent plant or plucked from it.

a) What does this tell you about the source of the growth regulator substance that triggers ripening?

b) Name this growth regulator substance.

21 The graph in figure 12.31 shows the respiration rate of avocado fruits over a period of two weeks after being picked.

a) Suggest why respiration rate fell during the first two days after the fruit was picked.

b) (i) On which day did respiration rate begin to increase?

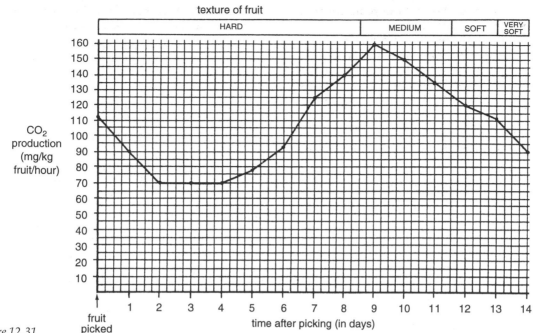

texture of fruit

| HARD | MEDIUM | SOFT | VERY SOFT |

CO₂ production (mg/kg fruit/hour)

fruit picked

time after picking (in days)

Figure 12.31

(ii) For how many days did this increase in respiration rate continue?

(iii) What effect did increase in respiration rate have on the texture of the avocado fruits?

c) (i) Name the growth substance that triggers the increase in respiration rate shown in the graph.

(ii) By what means does such an increase in respiration rate bring about the process of ripening?

d) (i) On which of the following days after picking would an avocado fruit probably be best to eat?

A 2   B 5   C 7   D 12 (Choose ONE answer only.)

(ii) Explain your choice of answer to (i) in terms of texture and flavour of fruit.

22 Figure 12.32 shows fully grown but unripe bananas about to be picked, boxed and shipped from a country in Central America. On arriving in Britain the bananas are still hard and green.

a) Describe one form of treatment that can be applied to ripen them before they are sent to the fruit market.

b) Briefly explain how the treatment works.

23 The graph in figure 12.33 shows the changes in concentration of two enzymes present in date fruit that occur before and during the process of ripening.

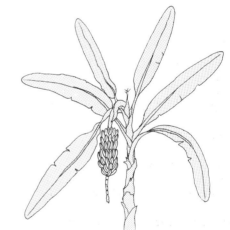

Figure 12.32

a) (i) Trace or redraw the graph and then add a curve to show the likely trend in ethylene production by date fruit during the period of ripening.

(ii) Explain the shape of your ethylene curve.

b) Which enzyme's action would affect (i) flavour (ii) texture of the dates?

c) Suggest a third way in which the dates would change between stages 1 and 5 of the ripening process.

159

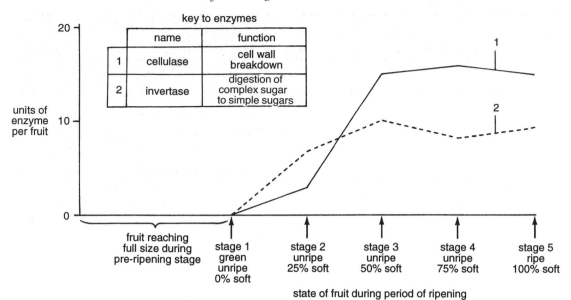

Figure 12.33

13 RESPONSES TO ENVIRONMENTAL STIMULI

TROPIC RESPONSES

COLEOPTILE

Figure 12.1 on page 142 shows an oat seedling with its first green leaves protected by a sheath called a **coleoptile.** Figure 12.2 shows several experiments involving coleoptiles and **auxin** (a type of plant growth substance).

GROWTH CURVATURE EFFECTS

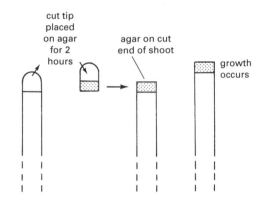

Figure 13.1: Use of agar block

The experiment in figure 13.1 shows that auxin will diffuse from a shoot tip into an agar block and then from the agar into a cut coleoptile where the cells resume elongation.

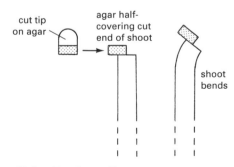

Figure 13.2: Bending effect

When an agar block containing auxin is placed 'half on and half off' the cut end of a coleoptile (see figure 13.2), the shoot **bends.** This happens because the side below the agar gets a **higher concentration** of growth substance causing greater cell elongation to occur on that side.

Within limits, the higher the concentration of auxin applied, the greater the amount of curvature produced.

TROPISM

A **tropic response** (tropism) is a growth movement of part of a plant in response to an external stimulus acting in **one direction.** The part of the plant affected grows either towards or away from the stimulus.

PHOTOTROPISM

This is the name given to a directional growth movement by a plant organ in response to **light** from one direction.

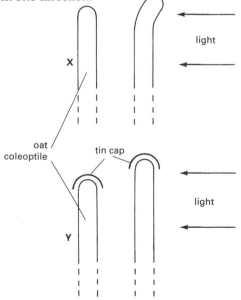

Figure 13.3: Response to light from one side

Oat coleoptiles exhibit **positive phototropism** by bending **towards** a unidirectional source of light as shown by shoot X in figure 13.3. Since shoot Y fails to bend, it is concluded that the shoot tip is the region responsible for detecting light acting on the plant from one side.

Mechanism of phototropism

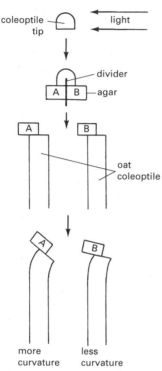

Figure 13.4:   Mechanism of phototropism

The experiment in figure 13.4 shows that a undirectional light source causes an **unequal** distribution of hormone to occur in the shoot tip. A higher concentration of **auxin** is present in the **non-illuminated** side than in the illuminated side.

These findings can be used to explain the mechanism of positive phototropism. If the shaded side of a shoot contains more auxin, then more cell elongation will occur on that side. This makes the shoot bend towards the light as shown in figure 13.5.

Importance of phototropism

Growth of a shoot towards light is of survival value to the plant because it is likely that the shoot will be exposed to the light energy necessary for photosynthesis.

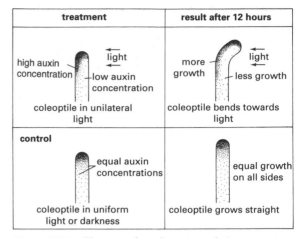

treatment	result after 12 hours
high auxin concentration — low auxin concentration — **light** coleoptile in unilateral light	more growth — less growth — **light** coleoptile bends towards light
control equal auxin concentrations — coleoptile in uniform light or darkness	equal growth on all sides — coleoptile grows straight

Figure 13.5:   Hormonal explanation of phototropism

KEY QUESTIONS

1 a) What is an auxin?
b) By what means can auxin be collected in an agar block?
c) An agar block containing auxin can be placed on the cut end of a coleoptile so the block covers either (i) the whole cut surface or (ii) only half of the cut surface. Describe the way in which a coleoptile would respond to each of these treatments.

2 a) What is meant by the term *tropic response?*
b) (i) Describe the movement of a shoot in response to light from one direction. (ii) Which side of such a shoot is found to contain a higher concentration of auxin? (iii) Which side of such a shoot is found to contain a lower concentration of auxin? (iv) Use these findings to explain the mechanism of phototropism.

3 What is the advantage to a plant of its shoots being positively phototropic?

GEOTROPISM

This is the name given to a directional growth movement by a plant organ in response to **gravity**.

The experiment illustrated in figure 13.6 shows the effect of gravity on the growth of a root and a shoot. (The apparatus is kept in darkness during the experiment to eliminate the effect of light.)

In the stationary clinostat, the root exhibits **positive geotropism** by growing downwards towards the source of the stimulus (gravity). The

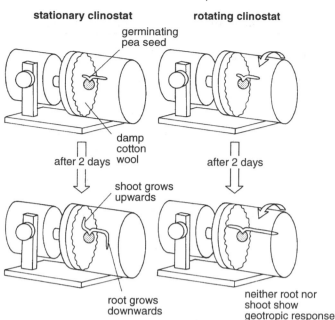

stationary clinostat   **rotating clinostat**

germinating
pea seed

after 2 days   damp
cotton
wool

after 2 days

shoot grows
upwards

root grows
downwards

neither root nor
shoot show
geotropic response

Figure 13.6:   Geotropism

shoot shows **negative geotropism** by growing upwards away from the pull of gravity.

In the rotating clinostat (the control experiment), gravity acts equally on all sides of each plant and therefore no geotropic responses occur.

Mechanism of geotropism

In a shoot, it is thought that a high concentration of **auxin** gathers and promotes cell elongation on the lower surface. This causes the shoot to bend and grow upwards, i.e. negative geotropism (see figure 13.7).

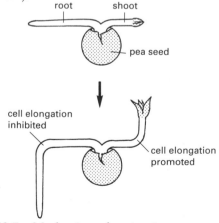

root   shoot

pea seed

cell elongation
inhibited

cell elongation
promoted

Figure 13.7:   Mechanism of geotropism

In a root it is thought that an **inhibitor** substance (**abscissic acid**) slows down cell elongation on the lower surface causing the root to bend and grow downwards i.e. positive geotropism.

Importance of geotropism

Growth of a root in a downwards direction is of survival value to the plant because this response is likely to make the root grow down into the ground giving it anchorage and ready access to supplies of water.

Growth of a shoot in an upwards direction is likely to give it access to the light energy needed for photosynthesis.

KEY QUESTIONS

1 a) What is meant by the term *geotropic response?*
  b) Which organ in a pea plant is (i) positively (ii) negatively geotropic?

2 a) (i) Which side of a shoot kept in a horizontal position is found to contain a higher concentration of auxin? (ii) Use this information to explain the mechanism of negative geotropism.
  b) (i) Which side of a root kept in a horizontal position is found to contain a higher concentration of the inhibitor, abscissic acid? (ii) Use this information to explain the mechanism of positive geotropism.

3 Of what advantage is it to a plant to have roots that grow downwards and shoots that grow upwards in response to gravity?

NASTIC RESPONSES

A **nastic response** is a movement by a plant in response to an external stimulus where the response is **non directional** in relation to the direction of the stimulus.

The external stimulus usually acts equally all round the plant e.g. change in light intensity or temperature.

LIGHT

The leaves of many leguminous (pod) plants such as clover are found to 'open' in light and keep their

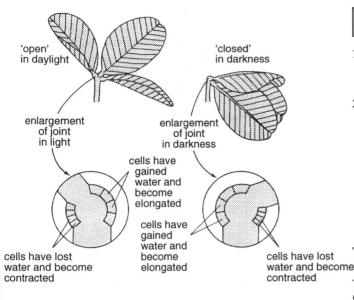

Figure 13.8: Photonasty in clover leaves

leaflets flat and expanded during the day as shown in figure 13.8. As night approaches and light gradually fades, the two side leaflets fold together and the third leaflet droops over them forming a temporary cover. This type of nastic response is also known as a 'sleep' movement.

Mechanism of response

Each leaflet of a leguminous plant capable of 'sleep' movements is found to be attached to the leaf stalk by a swollen **joint**. The movements of the leaflet are brought about by **differential changes** in the **turgor** of the cells on the upper and lower surfaces of the joint in response to an increase or decrease in light intensity.

On one side of the joint the cells rapidly lose water and become contracted, whilst on the other side the cells gain water and become elongated. These movements in turn cause the leaflets to move as shown in figure 13.8.

Survival value

This type of nastic response (called **photonasty**) enables the plant to present the maximum surface area of green leaf for light absorption during the day yet reduce the surface area of the leaf liable to water loss by transpiration at night.

KEY QUESTIONS

1 **a)** What is meant by the term *nastic response?*
   b) What is the basic difference between a **nastic** and a **tropic** response?

2 **a)** Describe the nastic movement shown by a clover leaf in response to (i) the arrival of daylight (ii) the onset of darkness.
   b) (i) Make a simple diagram of the point of attachment of a clover leaflet to its leaf stalk to show the effect of darkness on the state of the cells in this area. (ii) What is the survival value to the clover plant of 'sleep' movements by its leaves?

TEMPERATURE

The flowers of some plants such as crocus and tulip are found to open in temperatures of about 16°C and above. At temperatures below 16°C they close their petals as shown in figure 13.9. This nastic response is called thermonasty.

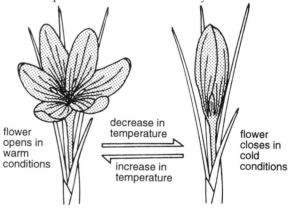

Figure 13.9: Thermonasty on crocus flower

Mechanism of thermonasty

Each petal capable of such nastic movements is found to possess an inner and an outer layer of cells which respond to temperature changes by **elongating at different rates**.

At temperatures above 16°C, the cells on the inner surface of the petals (which are especially sensitive to increase in temperature) elongate much more rapidly than those on the outside causing the petals to unfold and the flower to open as the sun warms it (see figure 13.10).

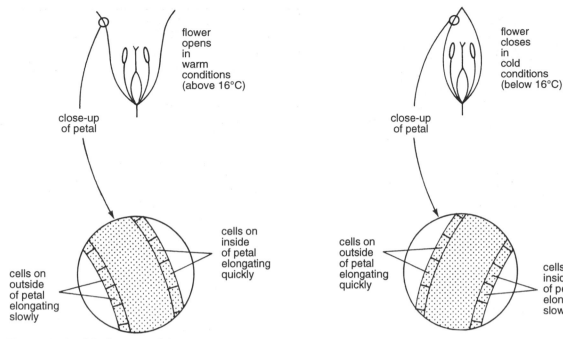

Figure 13.10: Mechanism of thermonasty in petals

At temperatures below 16°C, the rate of elongation of cells on the inner surface decreases to a level below that of the cells on the outside causing the petals to close.

Importance of thermonasty

This type of nastic movement is of survival value to the plant since it enables the flower to present its male and female reproductive parts in warm weather when pollinating insects are most active; the reverse response enables it to close and protect its reproductive organs from damage when the weather is cold.

TOUCH

Sundew (see figure 13.11) is an insectivorous plant which grows in swamps and acidic bogs in Britain. Each of its leaves bears about 200 tentacles which are used to trap insects.

An insect which lands on a leaf becomes caught up in sticky 'dew' produced by the tips of the tentacles. This is followed by a general **bending** of the tentacles towards the centre of the leaf which results in the insect becoming trapped. The tips of the tentacles then secrete an enzyme which digests protein present in the insect's body. Once the end

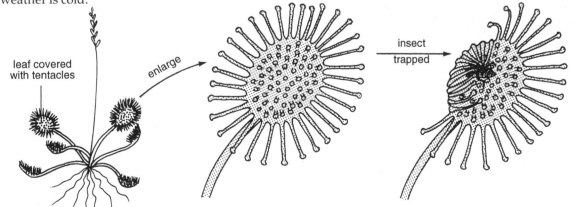

Figure 13.11: Sundew

products of digestion have been absorbed by the leaf cells, the tentacles return to their original position and the undigested remains of the insect blow away.

Mechanism of the response

The movement of sundew's tentacles is a further example of a nastic response. In this case the stimulus is **touch.** Bending of each tentacle is brought about by an **increase in size** of the cells on the lower side of the tentacle near its base as shown in figure 13.12. Return of a tentacle to its normal extended position is the result of increase in size of the cells on the upper surface of the tentacle.

Importance

This type of nastic response enables sundew to capture prey containing a supply of **nitrogen** (in protein). It is of survival value because it provides the plant with nitrogen which is scarce or lacking in poor infertile acidic bog soil.

KEY QUESTIONS

1 a) Describe the nastic movement shown by the petals of a crocus flower in response to (i) a decrease in temperature to below 16°C (ii) an increase in temperature to above 16°C.
b) (i) Make a simple diagram of part of a crocus petal to show the state of the cells of its inner and outer layers at 20°C. (ii) Which layer of cells elongates at the faster rate in these warm conditions? (iii) Which layer of cells would elongate at the faster rate if the flower were placed in a room at 7°C?

2 a) Such movements in crocus flowers are described as thermonastic. Suggest why.
b) If temperature is kept constant, crocus flowers will open in light and close in darkness.

What would be an appropriate name for this type of nastic response?

3 Of what advantage are such nastic movements to a crocus plant?

4 a) What is meant by the term *insectivorous* when applied to plants?
b) How does a sundew plant capture its prey?
c) Describe at cellular level the means by which one tentacle is moved.
d) Of what advantage to a sundew plant is this nastic response?

PHYTOCHROME-MEDIATED RESPONSES

In addition to photosynthesis, light plays a key role in many other aspects of a plant's growth and development. These include the ways in which certain parts of some plants respond to **red** light and **far red** light. (Red light has a wavelength of 660 nm and is part of the visible spectrum; far red light has a wavelength of 730 nm and is not part of the visible spectrum.)

The seeds of some varieties of lettuce will only germinate if exposed to light. Red light is found to be the most effective at breaking their dormancy; far red light has the reverse effect and inhibits germination.

Examples of responses by plant parts to red and far red light are given in table 13.1.

aspect of growth and development	type of light	
	red	far red
germination of seeds of some varieties of lettuce	stimulation	inhibition
stem elongation	inhibition	stimulation
leaf expansion	stimulation	inhibition

Table 13.1   Responses to red and far red light

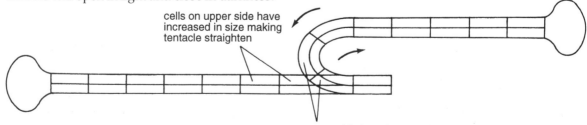

cells on upper side have increased in size making tentacle straighten

cells on lower side have increased in size making tentacle bend

Figure 13.12:   Mechanism of tentacle movement

Phytochrome

To be able to respond to red and far red light, a plant must first possess a photoreceptor that absorbs these wavelengths of light. The photoreceptor is a substance called **phytochrome** which exists in two forms.

Phytochrome 660 (P_{660}) absorbs red light (with an absorption peak at a wavelength of 660 nm); phytochrome 730 (P_{730}) absorbs far red light (with an absorption peak at a wavelength of 730 nm) as shown in the graph in figure 13.13.

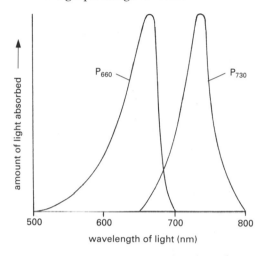

Figure 13.13: Absorption spectra for phytochromes

On absorbing is particular wavelength of light, each form of phytochrome becomes converted into the other form (see figure 13.14.)

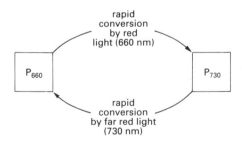

Figure 13.14: Interconversion of phytochrome

Since sunlight contains much more light of wavelength 660 nm than 730 nm, during daylight P_{660} becomes converted to P_{730} which then accumulates. P_{730} is unstable and during darkness slowly changes back into P_{660} which then accumulates as shown in figure 13.15.

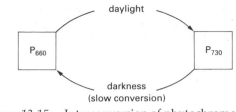

Figure 13.15: Interconversion of phytochrome

Mediation of response

It is thought that phytochrome becomes activiated on absorbing the appropriate wavelength of light. In the case of the lettuce seeds referred to above, P_{730} which accumulates during daylight hours is the active form. It causes one or more **genes** to be switched on. This results in the production of an active form of a hormone (probably **gibberellic acid**) from an inactive form.

The hormone then breaks the seed's dormancy and germination begins. This is called a **phytochrome-mediated response.**

Biological importance

It is possible that the light requirement for lettuce seeds to germinate is of survival value in that they will only start to grow at the surface of an 'open' stretch of soil. This helps them to avoid competitors which would block out light during their early growth.

Key Questions

1 a) What colour is light of wavelength 660 nm?
 b) What effect does this type of light have on
 (i) stem elongation (ii) leaf expansion in plants?

2 Seed germination of some types of lettuce can be stimulated using red light. By what means can germination of these seeds be inhibited?

3 a) What is phytochrome?
 b) Name the TWO forms of phytochrome and the type of light that each absorbs.
 c) Draw a simple diagram to show the interchangeable relationship that exists between the two forms as each absorbs its particular type of light.

4 a) Which form of phytochrome accumulates in the types of lettuce seed referred to in question 2 after they have been in daylight for a long period?

b) Suggest how this leads to the breaking of seed dormancy.

PHOTOPERIODISM

The number of hours of light in every 24 hours to which a plant is exposed is called the **photoperiod.**

EFFECT OF PHOTOPERIOD ON FLOWERING

Some plants respond to a change in photoperiod by stopping the production of vegetative (leaf) buds and starting the formation of flower buds. Such a response to a photoperiod is called **photoperiodism.** Three distinct categories of flowering plant are known to exist.

Long day (short night) plants only flower when the number of hours of light to which they are exposed is **above** a certain critical level (i.e. the number of hours of darkness is below the critical level). Spinach, for example, must receive at least

13 hours of light in order to flower (see figure 13.16). The length of the critical period of light varies amongst different species of long day (short night) plants (see table 13.2). In some cases it is found to be less than 12 hours.

Short day (long night) plants only flower when the number of hours of light to which they are exposed is **below** a certain level (i.e. the number of hours of darkness is above the critical level). Strawberry, for example, must receive at least 14 hours of darkness in order to flower (see figure 13.7.) The length of the critical period of darkness varies amongst different species of short day (long night) plants (see table 13.2). In some cases it is found to be less than 12 hours.

Day neutral plants are those in which flowering is not dependent upon photoperiod. Examples include celery, geranium and tomato.

Mechanism of response

Again the photoreceptor substance is phytochrome which exists as the two interconvertible forms described on page 164.

Florigen

Flowering in plants is thought to be initiated by a **hormone** (or combination of hormones). Since this hormone has not yet been isolated, it is provisionally referred to as **'florigen'.**

species of long day (short night) plant	critical duration of light (hours)	species of short day (long night) plant	critical duration of dark (hours)
Dill	11	Bryophyllum	12
Italian ryegrass	11	Chrysanthemum	9
Red clover	12	Cocklebur	9
Spinach	13	Strawberry	14
Winter wheat	12	Winter rye	12

Table 13.2   Critical periods of light and darkness

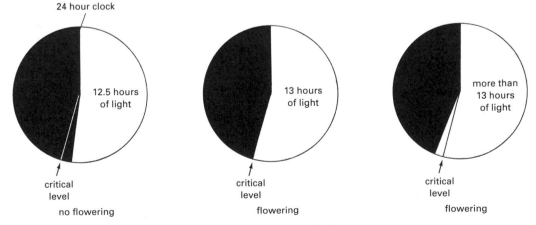

Figure 13.16:   Photoperiodism in spinach, a long day (short night) plant

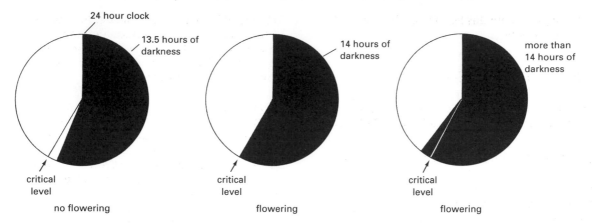

Figure 13.17:   Photoperiodism in strawberry, a short day (long night) plant

In long day plants, a high concentration of P_{730} is required for the release of florigen (see figure 13.18).

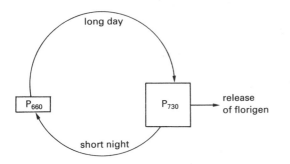

Figure 13.18: P_{730} formation in long day plant

In short day plants, a high concentration of P_{660} is required for the release of florigen (see figure 13.19). If the long night in interrupted somewhere near its middle by a flash of red light, flowering does not occur because some of the essential P_{660} has changed back into P_{730}.

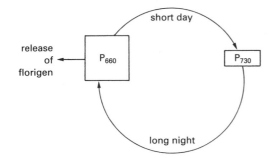

Figure 13.19: P_{660} formation in short day plant

The experiment shown in figure 13.20 shows that it is the **leaf** that detects the stimulus and responds by producing sufficient florigen to induce flowering in the entire plant.

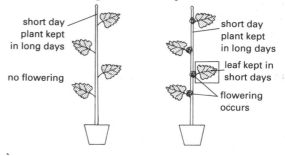

The experiment in figure 13.21 shows that florigen is **translocated** in the **phloem** to growing points where flowering is induced by the switching on of certain genes.

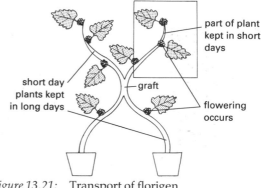

Figure 13.21:   Transport of florigen

Thus flowering in long day and short day plants is both a **phytochrome-mediated** and a **photoperiodic response.**

Biological importance

By responding to a photoperiod of particular length, all members of a species produce their flowers at the same time of year. This allows **cross-pollination** to occur, often on a massive scale.

Cross-pollination is of advantage to the species because it results in the production of much **variation** amongst the offspring and therefore increases the species' long-term chance of survival (see also page 00).

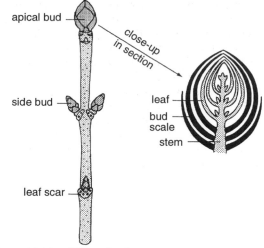

Figure 13.22: Winter buds

Buds develop on the young branches of deciduous trees during the growing season in summer. The buds become dormant in autumn and remain in this inactive state throughout winter. When spring arrives, the dormancy is broken and the buds grow into new leafy shoots.

MEDIATION OF RESPONSE

The stimulus for the onset of bud dormancy is the series of **decreasing photoperiods** that occurs in autumn in temperate climates. These are detected by **phytochrome** which is present in the bud scales. Dormancy occurs only if long dark periods are available to allow P_{730} to change into P_{660}. (Dormancy can be prevented by interrupting the long night with flashes of red light.)

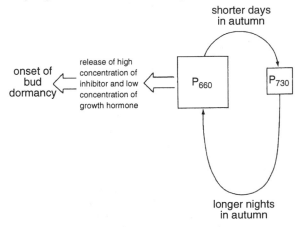

Figure 13.23: Events leading to onset of bud dormancy

KEY QUESTIONS

1 Define the term *photoperiod* with respect to a plant.

2 **a)** Explain the difference between long day and short day plants.
   b) Give a named example of each type and state its critical photoperiodic requirement for flowering.

3 **a)** (i) Which type of light is absorbed by phytochrome 660 (P_{660})? (ii) Which type of light is absorbed by phytochrome 730 (P_{730})?
   b) Which form of phytochrome accumulates during (i) long days (short nights) (ii) short days (long nights).
   c) (i) Accumulation of the appropriate form of phytochrome in a plant triggers the release of one or more hormones. What provisional name is given to the hormone(s)? (ii) What effect does this hormone have on a growing point?

4 Flowering in certain species of plant is a photoperiodic response. What possible advantage is gained by a plant species responding in this way?

EFFECT OF PHOTOPERIOD ON PERENNATING STRUCTURES

A perennating structure is an organ which enables a plant to survive the winter. Examples include winter buds on deciduous trees and underground tubers of potato plants.

DORMANCY OF WINTER BUDS

A **bud** is a miniature shoot (see figure 13.22) with a very short stem and closely packed leaves. Its thick protective outermost leaves are called **bud scales.**

Accumulation of P_{660} (see figure 13.23) causes one or more **genes** to be switched on. This leads to an increase in concentration of growth **inhibitor** (probably abscissic acid) and a decrease in concentration of growth-promoting hormone (probably gibberellic acid).

In spring the reverse occurs in response to the stimulus of **increasing photoperiods.** A high concentration of P_{730} accumulates (see figure 13.24) and leads to a decrease in concentration of the inhibitor and an increase in concentration of the **growth-promoting hormone** which then breaks dormancy. (Application of gibberellic acid to dormant buds of many types of deciduous trees during winter is found to break their dormancy artificially.)

The onset of dormancy and the breaking of dormancy are further examples of responses which are both **photoperiodic** and **phytochrome-mediated.**

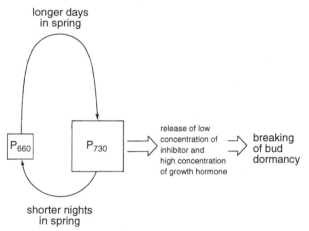

Figure 13.24: Events leading to breaking of bud dormancy

Survival value

The advantage provided by winter buds is twofold.

- By becoming dormant in response to photoperiods of decreasing length, they prevent the tree from running the risk of having its young shoots killed by low temperatures or lack of soil water during the icy winter months.
- By opening in response to photoperiods of increasing length, each bud provides the tree with a ready-made set of leaves for photosynthesis. This avoids the delay of several weeks that leaf formation would take and allows the tree to resume rapid growth in the favourable conditions of spring.

POTATO TUBERS

Some varieties of potato plant form **tubers** in response to photoperiods of decreasing length. The leaves are the site of the photoperiodic reaction involving phytochrome. Hormones are then translocated to the site of tuber formation.

KEY QUESTIONS

1 a) At about what time of the year do buds become dormant on deciduous trees?
b) To what stimulus does the tree respond by making its buds become dormant?
c) Which part of the plant contains the photoreceptor?

2 a) Name the photoreceptor referred to in question 1.
b) Which form of this substance accumulates at the end of winter in response to photoperiods of increasing length?
c) (i) What effect does a build-up of this form of the photoreceptor have on the hormonal balance present in the dormant buds? (ii) What do the buds do now? (iii) At what time of the year does this happen? (iv) Why is it of advantage to the plant that the response occurs at that time of year?

EXERCISES

1 Match the terms in list **X** with their descriptions in list **Y**.

list X	list Y
1) tropic response	a) form of photoreceptor that accumulates following a period of darkness or far red light
2) geotropism	b) general name for movement of a plant that is independent of the direction in which the external stimulus is acting on the plant
3) photo-tropism	c) form of photoreceptor that accumulates following a period of daylight or red light

4) nastic response

5) photonasty

6) thermo-nasty

7) phyto-chrome-mediated response

8) phyto-chrome 660

9) phyto-chrome 730

10) photo-periodic response

d) tropic movement in response to light

e) type of phytochrome-mediated response which depends on length of periods of daylight to which plant is exposed

f) general name for stimulation or inhibition of growth of a plant organ following exposure to red or far red light detected by phytochrome

g) nastic movement in response to light

h) general name for movement of part of a plant that is dependent on the direction in which the stimulus is acting on the plant

i) tropic movement in response to gravity

j) nastic movement in response to temperature

2 Which ONE of the sets of apparatus shown in figure 13.25 could be used to find out if the root of a pea seedling is positively geotropic?

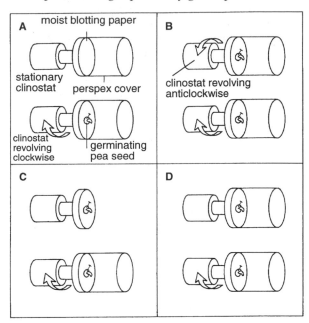

Figure 13.25

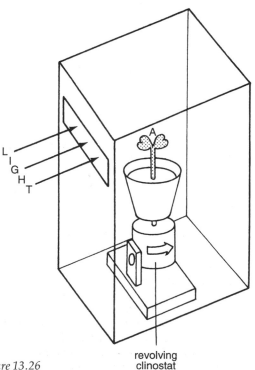

Figure 13.26

3 Figure 13.26 shows the start of an experiment set up to test radish seedlings using clinostats.
a) (i) Which plant will have received light on only one side of its shoot throughout the experiment?
(ii) Which plant will have received an equal amount of light on all sides of its shoot during the experiment?
(iii) Redraw plants A and B to show their appearance after two days of the treatment.
b) (i) Which plant would show a phototropic response after two days?
(ii) Would this response be positive or negative phototropism?
c) When a shoot is illuminated from one side, a difference in concentration of growth hormone is found to occur between the light and the dark sides of the shoot. State which side contains more hormone and explain how this results in the shoot bending.
d) Why is it of advantage to the plant to have a shoot that bends in this way?

4 Figure 13.27 shows an experiment set up to test young shoots using clinostats in a dark cupboard.

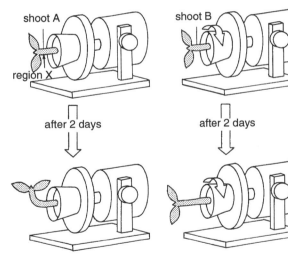

Figure 13.27

a) Name the type of response shown by shoot A.
b) Identify the environmental stimulus to which shoot A responded.
c) (i) Which type of chemical gathers at region X in shoot A?

(ii) Why does a high concentration of this chemical not gather in the lower side of shoot B (the control)?
(iii) Explain the bending of shoot A in terms of the effect the chemical has on its cells.
d) Of what advantage is it to a plant to have a shoot that responds to an environmental stimulus by growing upwards?

5 Figure 13.28 shows six stages involved in constructing and testing a hypothesis. Arrange the stages into the correct order starting with **F**.

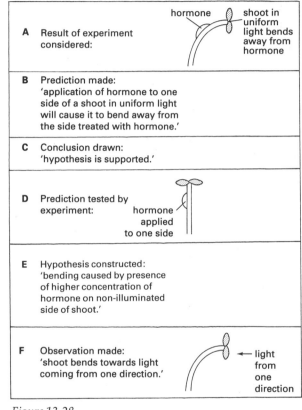

Figure 13.28

6 The experiment shown in figure 13.29 is set up to investigate tropic responses made by the young root and shoot of a bean seedling.
a) (i) Predict the direction in which the young root will grow.
(ii) Explain your answer.
b) The young shoot is found to grow vertically upwards. A pupil concludes that the shoot is showing a negatively geotropic response. Why isn't she justified in drawing this conclusion?

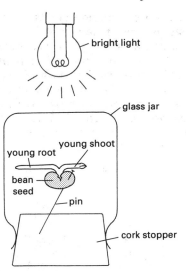

Figure 13.29

c) In what way would the experiment have to be altered to find out if the shoot is able to make a negatively geotropic response?

Exercises 7, 8 and 9 are multiple choice items. In each case you should choose ONE correct answer only.

7 Gardeners do not have to worry about planting seeds upside down because roots, on emerging at germination, show
  A negative phototropism.
  B positive phototropism.
  C negative geotropism.
  D positive geotropism.

8 Figure 13.30 shows four coleoptiles set up at the start of an experiment.

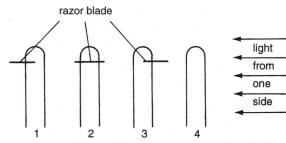

Figure 13.30

Which two coleptiles will BOTH bend towards the light source?
  A 1 and 2
  B 1 and 4
  C 2 and 3
  D 3 and 4

9 Tubes of cress seedlings were set up as shown in figure 13.31.

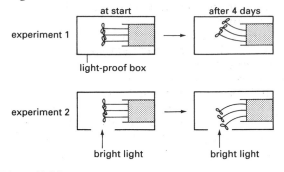

Figure 13.31

From the results of these experiments it can be concluded that
  A negative phototropism is a more powerful response than positive geotropism.
  B positive phototropism is a more powerful response than negative geotropism.
  C positive geotropism is a more powerful response than negative phototropism.
  D negative geotropism is a more powerful response than positive phototropism.

10 Figure 13.32 shows two stages in the life of a poppy flower. Region X of the stem bearing the young floral bud is found to be positively geotropic until the flower is ready to open.

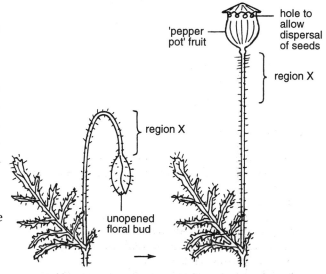

Figure 13.32

At a later stage, during seed and fruit development, the stem, now bearing the 'pepper-pot', is found to be negatively geotropic.

a) (i) What is meant by the term geotropism?
(ii) To what stimulus is geotropism the response?
(iii) What is the difference between positive and negative geotropism?
b) Suggest the survival value to a poppy plant of having (i) an unopened floral bud that droops (ii) a fully formed fruit that is held up in an erect position.

11 Figure 13.33 shows an experiment using pea seeds. It was carried out in a dark cupboard at 20°C with the apparatus held at the angle shown in the diagram throughout the experiment.

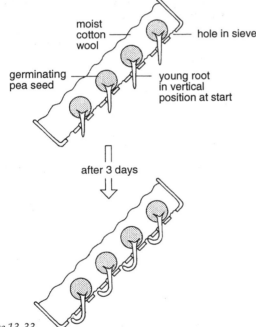

Figure 13.33

a) Name the type of response shown by the pea roots from the following possible answers:
A tropic
B nastic
C photoperiodic
D phytochrome-mediated
b) Identify the environmental stimulus to which the roots responded.
c) What is the survival value to pea seedlings of having roots that respond in this way?

12 Read the passage below and then answer the questions which follow it.
A tendril is a plant organ which is sensitive to contact. When the tendril of a climbing plant comes in contact with a solid object (e.g. wooden cane), the cells on the contact side of the tendril elongate more slowly than the cells on the non contact side. This results in bending and coiling of the tendril around the object as shown in the figure 13.34.

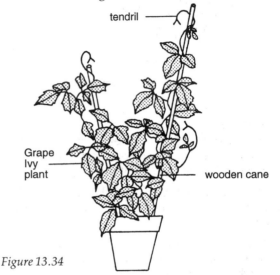

Figure 13.34

Under favourable conditions, the tendril may begin to curve within a few minutes of being touched on one side. This tropic response is known as thigmotropism.
a) To what stimulus is a tendril responding when it moves by thigmotropism?
b) (i) From the passage describe a thigmotropic response that can be observed without the aid of a microscope.
(ii) Describe how the plant brings about this response.
c) Suggest why thigmotropism is of survival value to Grape Ivy plants.

13 Figure 13.35 summarises an investigation into the effects of environmental stimuli on the opening and closing of dandelion 'flowers' (inflorescences).
a) (i) Which flowers responded by opening?
(ii) Identify the stimulus to which they responded.
b) (i) Which flowers responded by closing?
(ii) Identify the stimulus to which they responded.

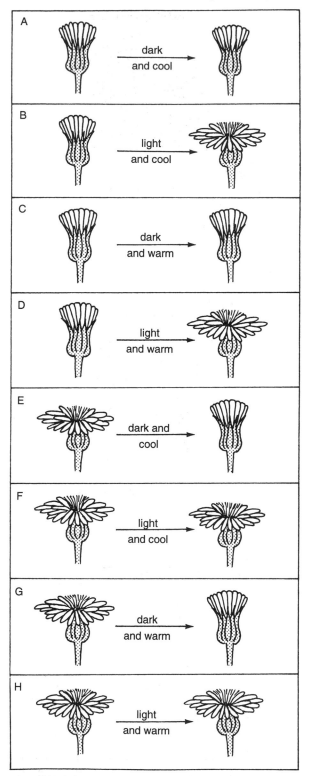

Figure 13.35

c) Suggest the survival value to dandelion plants of their flowers responding to the environmental stimulus in this manner.

d) (i) What general name is given to the type of plant response shown by the dandelion flowers?

(ii) In what way is this type of response different from a tropic response?

14 The graphs in figure 13.36 summarise the results of an investigation into the effects of change in temperature on the growth of strips of cells from the inner and outer surfaces of tulip flowers.

In experiment 1, the plant material was kept at 7°C for the first hour and then the temperature was quickly increased to 20°C (as indicated by the arrow).

In experiment 2, the plant material was kept at 20°C for the first hour and then the temperature was quickly decreased to 7°C (as indicated by the arrow).

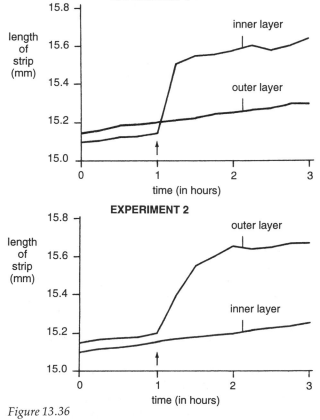

Figure 13.36

a) Which of the cell layers showed a steady increase in length *before* the temperature was changed in experiment 1?

b) (i) Which cell layer's rate of increase in length rose rapidly after the temperature change in experiment 1?

(ii) Which cell layer's rate of increase in length remained unaffected by temperature change in experiment 1?

(iii) What effect would such differential growth have on the tulip flower as a whole?

c) Which of the cell layers showed a steady increase in length *before* the temperature was changed in experiment 2?

d) (i) Which cell layer's rate of increase in length remained unaffected by temperature change in experiment 2?

(ii) Which cell layer's rate of increase in length rose rapidly after the temperature change in experiment 2?

(iii) What effect would such differential growth caused by a decrease in temperature have on the tulip flower as a whole?

(iv) What general name is given to this type of response?

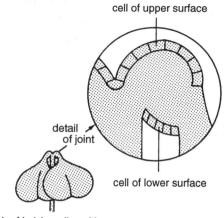

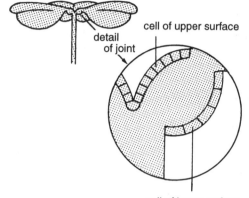

15 Each of the three leaflets of a wood sorrel leaf is connected to the leaf stalk by a joint. The water concentration of the cells in the upper and lower surface of each joint varies depending on the time of day as shown in the graph in figure 13.37. These changes in turn determine whether the leaf is 'closed' or 'open' as shown in the diagram.

a) (i) During which two-hour time interval did the water concentration of the cells in the upper side of the joint increase and those in the lower side decrease?

(ii) Which set of cells becomes elongated as a result?

(iii) What effect does this have on the overall appearance of the leaf?

b) (i) During which two-hour time interval did the water concentration of the cells in the lower side of the joint increase and those in the upper side decrease?

(ii) Which set of cells becomes elongated as a result?

(iii) What effect does this have on the overall appearance of the leaf?

c) These changes in a wood sorrel leaf are found to occur whether the weather is warm or

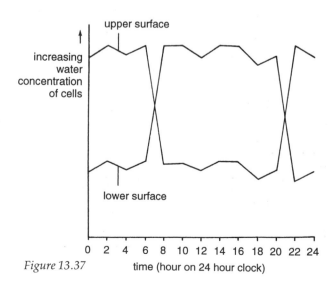

Figure 13.37

cold. Suggest which environmental stimulus brings about the response.

d) What is the proper name for this 'sleep movement' by the plant?

e) Of what survival value to the plant are these responses made by its leaves to an environmental stimulus?

16 Four groups of crocus plants were kept in darkness in a fridge for 24 hours and then treated as shown in figure 13.38.

a) (i) Which groups of crocus flowers responded by opening up?

(ii) Identify the stimuli which can bring about this response.

(iii) Are both stimuli necessary for the opening response? Explain your answer.

(iv) Name the type of plant response shown by these flowers when they open.

b) With reference to the cellular layers on the inside and outside of the petal, explain how the response is brought about.

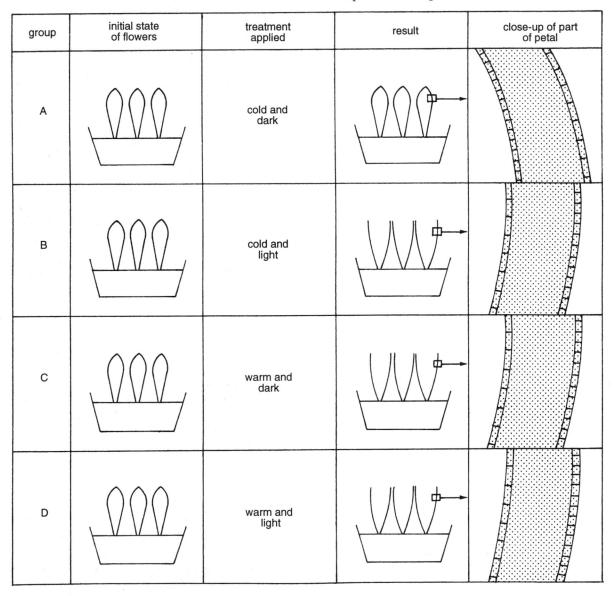

group	initial state of flowers	treatment applied	result	close-up of part of petal
A		cold and dark		
B		cold and light		
C		warm and dark		
D		warm and light		

Figure 13.38

c) Of what advantage is it to the crocus plant to open its flower in response to the stimuli that you identified?

17 An experiment using lettuce seeds of a certain variety was set up to investigate the factors affecting their germination. The results are summarised in table 13.3.

		seed coat intact	seed coat removed
type of light applied	red	+	+
	far red	−	+
	daylight	+	+

(+ = germination stimulated
− = germination inhibited)

Table 13.3

a) Identify the stimulus that normally leads to seed germination of this type of lettuce seed.
b) Name the photoreceptor substance which detects this stimulus.
c) (i) Where is the photoreceptor substance located in a lettuce seed?
(ii) Explain your answer.

18 During an investigation, lettuce seeds of a certain variety were exposed to various light stimuli. The results are summarised in the bar graph shown in figure 13.39.

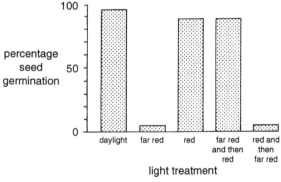

Figure 13.39

a) What control should have been run in the experiment in addition to the treatments referred to in the graph?
b) Which type of light (i) stimulates (ii) inhibits germination of this type of lettuce seed?
c) (i) What name is given to the photoreceptor represented by P in figure 13.40?

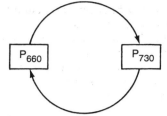

Figure 13.40

(ii) Which type of light makes P_{660} change into P_{730}?
(iii) Which type of light makes P_{730} change into P_{660}?
(iv) Which form of P must accumulate before stimulation of germination in these lettuce seeds occurs?
(v) Why does red light followed by far red light not stimulate germination?
d) What role is played out by the active form of P which eventually leads to seed germination?
e) Suggest the survival value to lettuce seeds of only being able to germinate in sites where light is available.

19 Table 13.4 shows the results of exposing three varieties of lettuce seed (X, Y and Z) to various light treatments. (R = red light, FR = far red light)

	% germination		
light treatment	variety X	variety Y	variety Z
R only	93	70	48
R then FR	20	74	0
R then FR then R	95	71	45
R then FR then R then FR	22	75	0
darkness	7	72	0

Table 13.4

a) Which variety did not depend on light to stimulate seed germination?
b) What effect did red light have on (i) variety X (ii) variety Y?
c) What effect did far red light have on (i) variety X (ii) variety Y?
d) (i) Predict the effect on each variety of applying the following light treatment: R then FR then R then FR then R then FR then R.
(ii) Give a reason for each of your predictions.

20 Figure 13.41 summarises photoperiodism in long and short day plants. Copy the diagram and complete the blanks using the following terms: daylight, release of florogen, P_{730}, darkness, production of flowers, P_{660}.

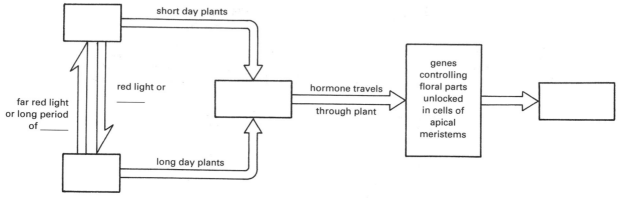

Figure 13.41

21 Chrysanthemums are short day plants which normally flower in autumn.
a) Which part of the plant must be exposed to the critical photoperiod for flowering to occur?
b) Describe the procedure that should be carried out by a commercial plant grower to produce a crop of Chrysanthemums in full bloom at Christmas time.

Exercises 22–27 are multiple choice items. In each case you should choose ONE correct answer only.

22 A short day plant will flower only when the continuous period of
A light is above a critical level.
B light is below a critical level.
C darkness is below a critical level.
D darkness is interrupted at a critical level.

23 Maryland Mammoth Tobacco is a short day plant. Its critical duration of darkness is ten hours. Under which of the conditions shown in figure 13.42 will it NOT flower?

24 The photoperiodic stimulus which leads to flowering is detected in a plant by the

A leaves.
B buds.
C flowers.
D shoot tips.

25 Phytochrome exist in two forms, P_{660} and P_{730}. Which of the conversions shown in figure 13.43 does NOT occur?

26 Which of the cycles shown in figure 13.44 correctly represents the events that occur in a short day plant?

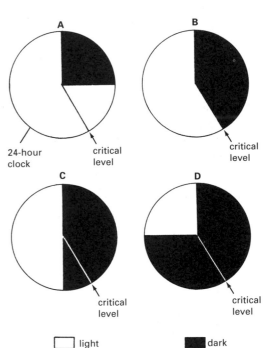

Figure 13.42

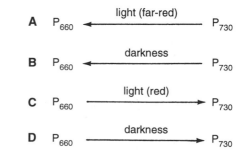

Figure 13.43

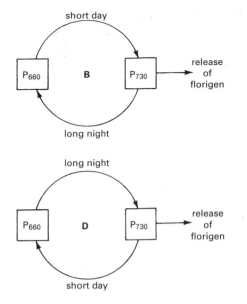

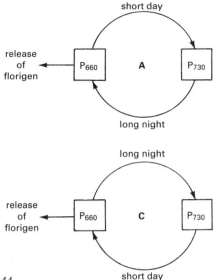

Figure 13.44

27 Cocklebur is a short day plant. After exposure to the treatments described in figure 13.45 for several days, plants X and Y were grafted together.

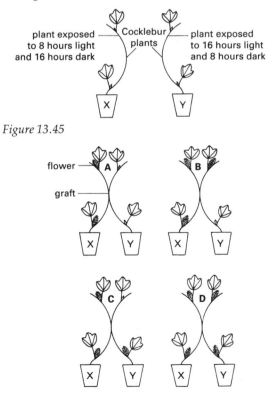

Figure 13.45

Figure 13.46

Which diagram in figure 13.46 correctly represents the appearance of the grafted plants when flowering occurs?

28 Read the passage below and then rewrite the sentences that follow it using only the correct answer from each choice.

Birch trees normally form dormant buds in autumn. In an experiment, a young birch tree grown in a greenhouse was given extra illumination as the lengths of the daily photoperiods began to decrease at the end of summer. The treatment was continued for six months. The young tree failed to make winter buds. Instead it maintained a state of continuous growth throughout the winter.

a) The normal environmental stimulus present in autumn consists of photoperiods of $\left\{ {\text{short} \atop \text{long}} \right\}$ days and $\left\{ {\text{short} \atop \text{long}} \right\}$ nights. However in this experiment the artificial stimulus consisted of a series of $\left\{ {\text{short} \atop \text{long}} \right\}$ days and $\left\{ {\text{short} \atop \text{long}} \right\}$ nights.

b) Under normal circumstances, long nights allow $\left\{ {P_{660} \atop P_{730}} \right\}$ to change to $\left\{ {P_{660} \atop P_{730}} \right\}$ which leads to the formation of $\left\{ {\text{an increased} \atop \text{a decreased}} \right\}$ concentration of inhibitor and $\left\{ {\text{an increased} \atop \text{a decreased}} \right\}$

concentration of growth-promoting hormone resulting in the buds becoming dormant.

c) However when extra light was given to the young birch tree, $\left\{\dfrac{P_{660}}{P_{730}}\right\}$ changed to $\left\{\dfrac{P_{660}}{P_{730}}\right\}$ and this active form of phytochrome led to the formation of $\left\{\begin{array}{l}\text{an increased}\\ \text{a decreased}\end{array}\right\}$ concentration of growth-promoting hormone resulting in the buds continuing to grow.

29 Begonia plants form tubers in response to daily photoperiods of 8 hours. An experiment was set up to investigate the importance of the 16-hour dark period to tuber formation. For several nights the dark period was interrupted by one hour of red light. Different groups of plants were given this light at different times during the 16-hour dark period. The bar graph in figure 13.47 summarises the results.

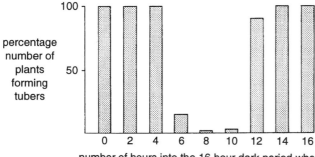

Figure 13.47

a) Identify the environmental stimulus that leads to the Begonia plants making tubers.
b) (i) Suggest the plant part which detects the stimulus.
(ii) Name the photoreceptor present in this plant part.
(iii) State the way in which this substance becomes altered during uninterrupted long nights.
(iv) Construct a flow diagram to show the possible series of events that leads to tuber formation.
c) (i) At which time during the 16-hour dark period is the interruption by one hour of red light most effective at inhibiting tuber formation?
(ii) State the effect that red light has on the balance between the two forms of the photoreceptor.

(iii) Suggest why this change in the balance prevents tuber formation.

30 Figure 13.48 shows an investigation into tuber formation in a certain variety of potato plant.

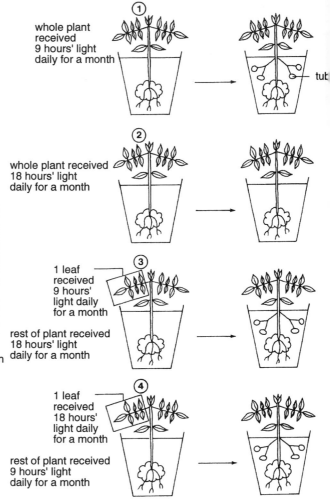

Figure 13.48

a) (i) Which plants responded to the light treatment that they received by forming tubers?
(ii) Identify the environmental stimulus to which they responded.
b) (i) Which part of the plant detected the stimulus?
(ii) What photoreceptor substance is present in this plant part?

(iii) Which form of the photorector must accumulate to lead to tuber formation?

(iv) Describe the chain of events that brings about this response.

c) To which TWO of the following categories of response does the one shown in this experiment belong?

A tropic

B nastic

C photoperiodic

D phytochrome-mediated

d) Of what advantage is it to potato plants to respond to the environmental stimulus by making tubers?

14 ADAPTATIONS TO ENVIRONMENTS

MESOPHYTES

A **mesophyte** is a plant which lives in an environment where there is an ample water supply. The majority of flowering plants that live on land in temperate climates are mesophytes.

A mesophyte is neither adapted to survive shortage of water (see xerophytes) nor adapted to survive being submerged in water (see hydrophytes). Mesophytes occupy an ecological niche **between** these two extremes. Water lost by a mesophyte during transpiration is readily replaced by uptake from the soil.

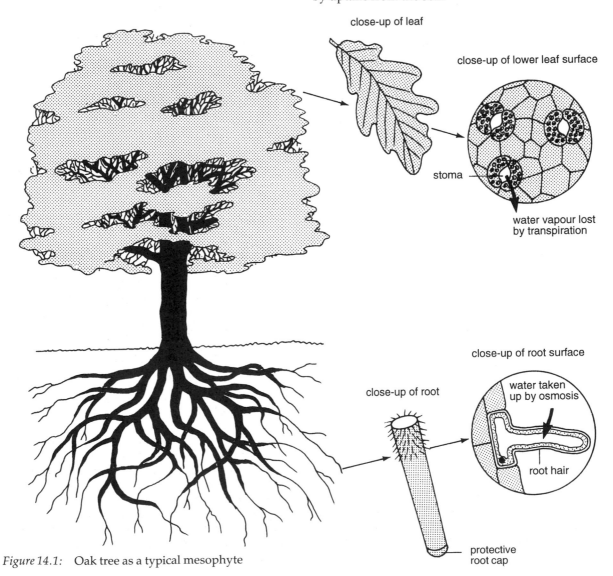

close-up of leaf

close-up of lower leaf surface

stoma

water vapour lost by transpiration

close-up of root surface

water taken up by osmosis

close-up of root

root hair

protective root cap

Figure 14.1: Oak tree as a typical mesophyte

A mesophyte's structure (see figure 14.1) perfectly suits it to life in its well-watered environment. It has **'normal' root hairs** for water absorption, a generous supply of **xylem** vessels throughout its organs for water transport and support, and a **large surface area** of thin green leaf tissue for photosyntheis. The leaves tend to lose a large amount of water as water vapour through **stomata** which are open in light and closed in darkness.

LEAF FALL

Winter is the one time of year when a mesophyte cannot depend on access to an adequate water supply since soil water is often in a frozen state for long periods. To survive this **physiological drought**, deciduous trees (e.g. oak) shed their leaves in autumn and therefore greatly reduce water loss by transpiration during the winter.

Annual plants overwinter inside **dormant seeds** and many perennial plants survive inside **storage organs** such as bulbs and corms.

KEY QUESTIONS

1 a) What is a mesophyte?
b) Give an example of a mesophyte.

2 a) Name the pores in a leaf through which water is lost by transpiration.
b) By what means does a mesophyte replace this lost water?

3 a) (i) At what time of year could a deciduous tree be threatened with physiological drought?
(ii) Explain why.
b) (i) What change is undergone by a deciduous tree which enables it to avoid this problem?
(ii) Explain why this change makes the tree able to avoid physiological drought.

XEROPHYTES

Xerophytes are plants which live in habitats where a mesophyte would not survive because its transpiration rate would be excessively high. Such habitats are characterised by either hot, dry conditions and lack of soil water (e.g. desert) or exposed, windy conditions (e.g. moorland).

Xerophytes are able to maintain a water balance in such extreme habitats because they possess the following adaptations.

STRUCTURAL ADAPTATIONS WHICH REDUCE TRANSPIRATION RATE

Water loss is cut to a minimum in leaves which possess a **reduced number of stomata** and are covered by a **thick cuticle**. Transpiration is further reduced if the leaf is **rolled** and/or **hairy** (see figure 14.2) since each of these adaptations traps a layer of moist, relatively immobile air between the stomata and the outer atmosphere.

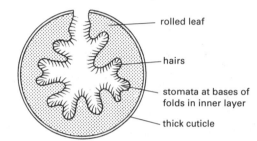

Figure 14.2:   Transverse section of marram grass leaf

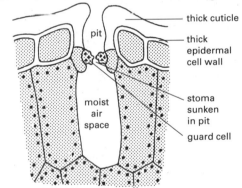

Figure 14.3:   Detail of *Hakea* leaf

Stomata sunken in **pits** (see figure 14.3) are similarly protected because the pits trap pockets of moist air.

Some leaves (e.g. pine needles) are **small** and **circular** in cross-section thereby reducing the relative surface area of transpiring leaf exposed to the atmosphere.

In many cacti such as *Echinocactus* (see figure 14.4), the leaves are reduced to protective **spines** and water loss is limited to the stem which carries out photosynthesis and possesses relatively few stomata.

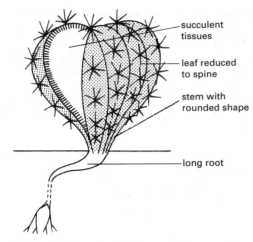

Figure 14.4: *Echinocactus* (with part of stem cut)

STRUCTURAL ADAPTATIONS FOR RESISTING DROUGHT

Many cacti have **long roots** allowing absorption of subterranean water. Others possess extensive systems of **superficial roots** which grow parallel to the soil surface enabling them to absorb maximum water on those rare occasions when rain does fall.

Cacti store this water in **succulent tissues** (see figure 14.4) and may even have a **folded stem** which allows expansion and contraction (subject to water availability) without becoming damaged.

PHYSIOLOGICAL ADAPTATIONS

Most cacti reduce water loss by showing **reversed stomatal rhythm** (i.e. closed during the day and open at night). During the night carbon dioxide is taken in and stored for use in photosynthesis in the daytime when the stomata are closed.

Some xerophytes evade drought by stopping normal growth during dry times. Some survive, for example, in a highly **dehydrated state** inside a hard seed coat and only germinate and grow when water becomes available.

KEY QUESTIONS

1 a) Describe TWO particular environments which are populated almost exclusively by xerophytes.
b) Why are mesophytes unable to survive in either of these habitats?

2 a) Give THREE examples of structural adaptations which reduce transpiration rate in xerophytes.
b) For each of these, describe how it brings about its effect.

3 a) Give THREE examples of structural adaptations which help a xerophyte to resist drought.
b) For each of these, describe how it plays its role.

HALOPHYTES

Halophytes are plants adapted to life in soil which is periodically swamped with sea water. Such soil is found in tidal mud flats and salt marshes at the coast.

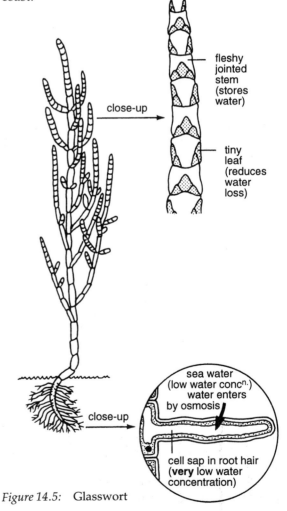

Figure 14.5: Glasswort

The water concentration of the cells of most land-living organisms is higher than that of sea water. However the root hair cells of a typical halophyte such as glasswort (see figure 14.5) contain cell sap which has a water concentration even **lower than that of sea water.** Halophytes are therefore able to gain water by osmosis from soil saturated with sea water.

The fleshy stems of glasswort and the fleshy leaves of sea plantain (another halophyte — see page 193) contain **succulent tissue** full of stored water. This is for use when evaporation from mud flats at low tide reduces the water concentration of the soil solution to a level lower than that of the root hairs. Under such conditions the plant loses water by osmosis to its surroundings until the soil is swamped by the next tide or a shower of rain falls.

KEY QUESTIONS

1 To which particular type of environment are halophytes adapted?

2 a) Give a named example of a halophyte.
b) What is unusual about this plant's root hairs?
c) In what way does this feature help the plant to survive in its habitat?

3 Where do halophytes store water for use on those occasions when they cannot gain water from their surroundings by osmosis?

HYDROPHYTES

Hydrophytes are plants which live completely submerged (e.g. water-milfoil) or partially submerged (e.g. water-lily) in water. They possess many adaptations which help them to survive in their aquatic environment.

AIR SPACES

Although gaseous exchange occurs all over its surface, a submerged plant is faced with the problem of obtaining an adequate supply of oxygen for respiration since this gas is only slightly soluble in water.

A hydrophyte is adapted to overcome this problem by possessing an extensive system of intercommunicating **air-filled cavities** (see figure 14.6) throughout its submerged organs. Instead of escaping into the surrounding water, much of the oxygen formed during photosynthesis is stored in these air spaces ready for use in respiration when required.

The presence of such **aeration tissue** also gives a submerged plant buoyancy keeping its leaves near the surface for light.

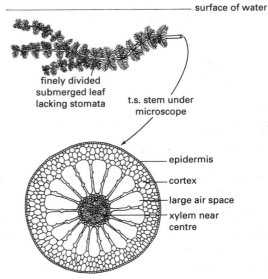

Figure 14.6:   Water-milfoil

REDUCTION OF XYLEM

Since water provides a submerged plant with support and is readily available for absorption all over its surface, a hydrophyte is found to possess little strengthening or water-conducting tissue.

Any xylem present is normally found at the centre of the stem (see figure 14.6). This allows the stem **maximum flexibility** in response to water movements while at the same time enabling it to resist pulling strains.

SPECIALISED LEAVES

A hydrophyte's submerged leaves are **narrow** in shape or **finely divided** (see figure 14.6). This adaptation helps to prevent them from being torn by water currents.

Stomata must be in contact with the air to bring about gaseous exchange. Submerged leaves therefore lack stomata and floating leaves have all

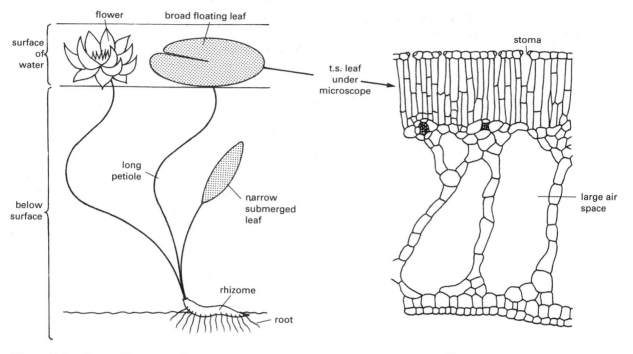

Figure 14.7: Water-lily

their stomata on their **upper surfaces** (see figure 14.7). Such floating leaves often have **long leaf stalks** which prevent the stomata from being flooded when the water level rises.

KEY QUESTIONS

1 a) What is a hydrophyte?
b) Including named examples in your answer, distinguish clearly between the two categories of hydrophyte.

2 a) Why is a submerged hydrophyte faced with the problem of obtaining an adequate supply of oxygen?
b) What structural adaptation does it have that enables it to solve the problem?
c) What further benefit is provided by the feature you gave as your answer to part b)?

3 a) Give TWO reasons why hydrophytes normally have very little xylem tissue compared with mesophytes.
b) (i) If a hydrophyte does have xylem, where is it located in its stem? (ii) Explain why.

Figure 14.8: Root tubers

4 a) Why is it of advantage to water-milfoil to have leaves that are finely divided?
b) (i) Where are stomata found on a floating leaf of water-lily? (ii) Explain why.

ADAPTATIONS FOR FOOD STORAGE

When a green plant produces more sugar by photosynthesis than it uses during respiration, the excess is stored (often as starch). Although food can be stored in the cells of almost any part of a plant, some species have organs especially adapted for this purpose.

MODIFIED ROOT

Dahlia plants form **root tubers** (see figure 14.8) during the growing season when food is plentiful.

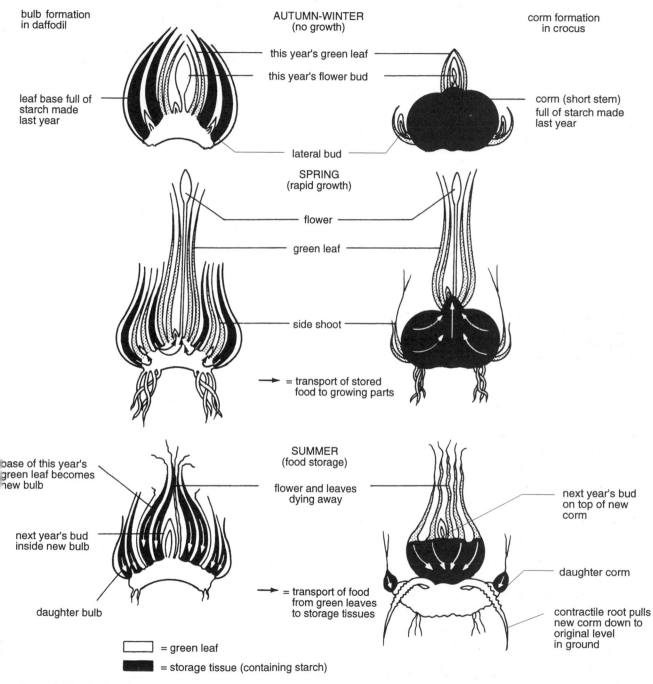

bulb formation in daffodil

AUTUMN-WINTER (no growth)

this year's green leaf

this year's flower bud

leaf base full of starch made last year

lateral bud

corm formation in crocus

corm (short stem) full of starch made last year

SPRING (rapid growth)

flower

green leaf

side shoot

→ = transport of stored food to growing parts

base of this year's green leaf becomes new bulb

SUMMER (food storage)

flower and leaves dying away

next year's bud inside new bulb

daughter bulb

→ = transport of food from green leaves to storage tissues

next year's bud on top of new corm

daughter corm

contractile root pulls new corm down to original level in ground

☐ = green leaf

■ = storage tissue (containing starch)

Figure 14.9: Bulbs and corms

189

The plant's leafy shoots die in autumn but the underground tubers, rich in stored food, remain alive throughout the winter in a dormant condition.

In spring new dahlia plants develop from the tuber cells using the stored food until they develop their own green leaves.

MODIFIED STEM

During the growing season, a crocus plant forms a swollen food-storing structure called a **corm** (see figure 14.9) which is a modified stem. Like a root tuber, the corm is the part of the plant which remains alive but dormant during the cold winter months.

In early spring rapid growth is resumed by the underground corm's apical bud. This develops into a shoot using the corm's starchy food reserves.

MODIFIED LEAF

The daffodil plant's organ of food storage is a **bulb** which consists of fleshy non-green leaves. These are really leaf bases which become swollen on receiving reserves of food from the upper green parts of the leaves during the growing season (see figure 14.9).

It is as an underground bulb that the daffodil survives the winter in a dormant state. The young shoot emerges in spring using the bulb's stored food for its rapid growth.

BIOLOGICAL ADVANTAGE

In addition to being adapted for storage of food, root tubers, corms and bulbs are the plants' **organs of perennation** (overwintering). The plant is able to survive the cold unfavourable winter conditions that occur in a temperate climate by 'ticking over' in a dormant state. Its cells operate at a very **low metabolic rate** and use hardly any of the stored food.

When spring arrives, the plant's lack of green leaves for photosynthesis is no handicap. Thanks to the food reserves in the cells of the modified organ, the young bud grows rapidly into a leafy shoot well ahead of many of its competitors.

COTYLEDONS

Annual plants die after one growing season. Their offspring survive the winter as seeds. The seeds of many annual plants contain two 'seed leaves' called **cotyledons** which are adapted for food storage.

Figure 14.10 shows a soaked broad bean seed opened up. Its cotyledons are found to be rich in starch.

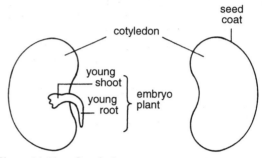

Figure 14.10: Cotyledons

In spring when warm moist conditions return after winter, the seeds of annual plants begin to germinate. Each seedling receives a supply of food from its cotyledons until it develops its own green leaves and becomes independent. The cotyledons also protect the embryo plant during the dormant period.

KEY QUESTIONS

1 a) Name a plant that makes root tubers.
   b) When does it produce them?
   c) At what time of year are the cells of a root tuber in a dormant state?

2 a) Give ONE structural difference between a corm and a bulb.
   b) State ONE feature that these organs have in common.
   c) Why do plants make modified organs such as corms and bulbs?

3 a) Describe how an annual plant's offspring manage to survive the rigours of a British winter.
   b) Why does the young root emerging from a germinating seed not die of starvation in the absence of green leaves for photosynthesis?

ADAPTATIONS AGAINST HIGH LEVELS OF INSOLATION

Insolation is the quantity of **solar radiation** falling on a plant during its exposure to the sun's rays. In Britain, plants rarely receive excessive amounts of sunshine. However in deserts and other extremely hot dry environments, exposure to intense solar radiation is the daily norm.

In the desert, it is the water supply that normally acts as the factor limiting the growth of the plant. Under such circumstances excess solar radiation is of little use to the plant photosynthetically; it may even prove to be hazardous by causing overheating and damage to the leaf.

Plants living in environments subjected to high levels of insolation have become adapted in several ways to prevent overheating.

LAYER OF INSULATION

Hairlike outgrowths are a common feature of many xerophytes. Figure 14.11 shows a cross section of a leaf from an Australian xerophyte which bears a coat of vesicular **'hairs'**. These trap a layer of air (a good insulator) and this helps to keep the heat out.

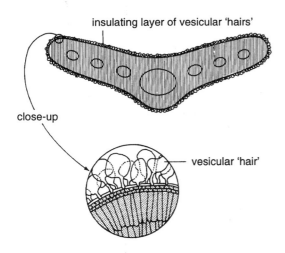

Figure 14.11:   Insulating layer of hairs in *Atriplex canescens*

The bark of most trees also acts as an effective layer of insulation against excessive heat.

REFLECTIVE LAYER

All leaves reflect some of the solar energy that falls on them, but compared with mesophytes, plants adapted to desert conditions reflect a far greater proportion of this radiation.

Many desert plants such as prickly pear (see figure 14.12) are made more reflective by the presence of an especially **shiny cuticle** on their aerial parts. The leaf blades of the eucalyptus tree (see figure 14.13) are covered with a layer of white **waxy material** which promotes the reflection of solar radiation.

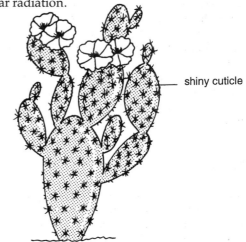

Figure 14.12:   Prickly pear

Some desert plants also possess a layer of **silvery hairs** which, in addition to providing insulation, help to reflect the sun's rays.

VERTICAL ORIENTATION OF LEAF

By hanging **vertically downwards**, the leaves of the eucalpytus tree (see figure 14.13) avoid direct exposure to the sun when it is directly above at the hottest part of the day.

NORTH-SOUTH ORIENTATION OF LEAF

Silphium laciniatum, the compass plant of the American prairies, turns its leaves edgeways due **north and south** (see figure 14.14 which shows a diagrammatic version of one leaf only).

By orientating its leaves in a north-south plane, the plant is able to avoid the full strength of the midday sun yet still make use of the less intense light falling on its leaves early and late in the day.

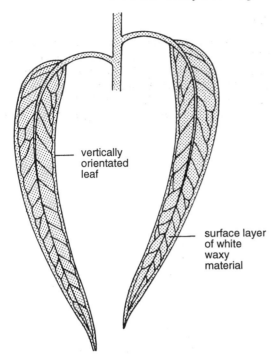

Figure 14.13: Vertical orientation of eucalyptus leaf

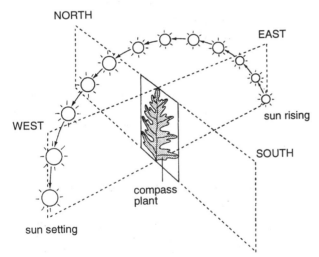

Figure 14.14: North-South orientation in compass plant

KEY QUESTIONS

1 What is meant by the term *insolation*?

2 Name a type of environment where the plants receive a high level of insolation on most days of the year.

3 Describe how the possession of hairlike outgrowths can adapt a plant to withstand high levels of insolation.

4 Describe how a waxy layer over its aerial parts can help a plant to survive the adverse effects of intense insolation.

5 Name a further adaptation against high levels of insolation and explain how it works.

ADAPTATIONS AGAINST PREDATION

Many plants are able to withstand predation by herbivorous animals. Some do this by producing **protective chemicals** in their tissues.

CYANIDE POISON

Hydrogen cyanide is a poison which acts by blocking an organism's cytochrome system (see page 39). Some varieties of white clover contain non-toxic glycoside which is hydrolysed by enzyme action to hydrogen cyanide. This only occurs when leaf tissue is damaged e.g. during nibbling by herbivores such as slugs. The process is summarised by the following equation:

$$\text{non-toxic glycoside} \xrightarrow{\text{enzyme action}} \text{toxic hydrogen cyanide}$$
following tissue damage

This production of cyanide is called **cyanogenesis.** Plants unable to make cyanide are said to be acyanogenic.

Table 14.1 shows the results of an experiment where several members of the same species of slug were released into plastic boxes containing equal numbers of leaves from cyanogenic and acyanogenic clover plants. The cyanogenic plants suffered much less damage.

		% number of leaves		
		undamaged	less than half of leaf eaten	half or more of leaf eaten
type of clover leaf	**cyanogenic**	80	16	4
	acyanogenic	42	25	33

Table 14.1   Damage to clover leaves

STRUCTURAL DEFENCE MECHANISMS

Amongst flowering plants, a variety of **structural** adaptations are found, each designed to keep hungry 'predators' at bay.

Thorns

Hawthorn trees bear sharp **thorns** which are modified side branches (see figure 14.15).

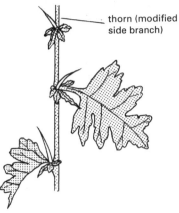

Figure 14.15:   Thorn

Spines

In gorse, the leaves are reduced entirely to short **spines** (figure 14.16) which are borne on modified branch stems. This presents an almost impenetrable barrier to hungry herbivorous mammals.

The leaves of some cacti (see figure 14.4) are completly reduced to spines. This helps to protect their inner succulent tissues from thirsty desert animals. In holly, the spines are restricted to the edges of the leaves.

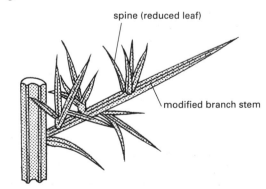

Figure 14.16:   Spine

Stings

The leaves of nettle possess **stinging hairs** (figure 14.17). Each takes the form of a thin capillary tube ending in a spherical tip. When an animal touches a hair, its spherical tip breaks off leaving a sharp edge. This penetrates the animal's skin allowing liquid irritant to be injected, giving an impressive warning.

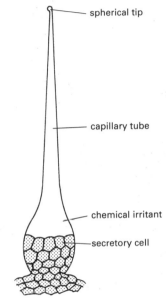

Figure 14.17:   Sting

KEY QUESTIONS

1 **a)** What is meant by the term *cyanogenesis?*
  b) Name a plant capable of employing this method of defence.
  c) Describe how the method works without poisoning the plant.

2 **a)** (i) List THREE different structural adaptations used by plants to repel hungry animals. (ii) Put in brackets after each of these an example of a plant that possesses the structure.
  b) Choose ONE of these adaptations and describe how it works as a defence mechanism.

ADAPTATIONS TO LOW-OXYGEN HABITATS

Mangrove trees are tropical evergreens that grow in swamps where the soil is periodically flooded

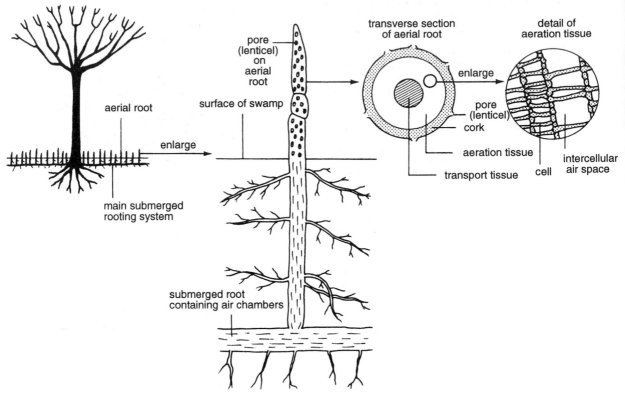

Figure 14.18: Aerial roots of mangrove tree

with sea water. Waterlogged soil contains very little oxygen since oxygen is only slightly soluble in water.

Mangrove trees are adapted to overcome this problem by possessing peg-like **aerial roots** which are negatively geotropic (see page 160). These develop from the main submerged rooting system and grow upwards sticking out of the stagnant swampy water as shown in figure 14.18.

In transverse section, an aerial root is found to possess transport tissue at its centre. This is surrounded by a wide layer of **aeration tissue** containing large intercellular **air spaces.** These connect with an extensive system of elongated **air chambers** present in the main submerged rooting system.

Each peg-like aerial root is covered with cork tissue containing numerous **pores** (lenticels) which allow gaseous exchange to occur. The aerial roots are often referred to as 'breathing' roots since they act as ventilators allowing oxygen to enter and carbon dioxide to leave by diffusion.

The system of interconnecting air chambers throughout the rooting system in turn allows this oxygen to diffuse to all parts of the root (most of which is submerged in anaerobic mud). It is by this means that the aerial roots adapt a mangrove tree to its low-oxygen environment.

KEY QUESTIONS

1 **a)** Describe the natural environment of a mangrove's roots.
b) V, hy does the soil in such a habitat contain very little oxygen?

2 State ONE way in which a transverse section (T.S.) of a mangrove tree's main root would differ from that of an oak tree.

3 **a)** State ONE behavioural, and TWO structural differences between an aerial root of a mangrove tree and a root of a germinating broad bean seed.
b) Describe the survival value to the mangrove tree of each of these three differences.

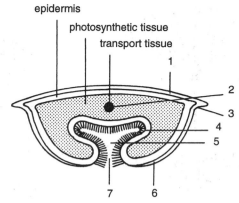

Figure 14.19

EXERCISES

1 In table 14.2, the information in columns X, Y and Z is mixed up. Copy the table and rearrange it so that the information in columns X, Y and Z correctly corresponds to the terms in column W.

W	X	Y	Z
type of plant	named example of plant type	potential problem caused by environment	adaptive feature of plant that overcomes problem
mesophyte	glasswort	lack of adequate oxygen for cell respiration	reduced number of stomata sunken in pits
xerophyte	oak	loss of cell water to salty soil solution	air-filled cavities for storage of oxygen made during photosynthesis
halophyte	water-milfoil	excessive loss of water by transpiration	shedding of leaves in autumn
hydrophyte	cactus	no problem except chance of physiological drought in winter	root cell sap with water concentration lower than that of sea water

Table 14.2

2 'On a sliding scale of water availability in their environments, mesophytes occupy a position intermediate between xerophytes and hydrophytes'.

a) State whether you agree or disagree.
b) Justify the choice you made in a).

3 Deciduous trees shed their leaves in autumn. Write a short paragraph to explain why this process is a physiological necessity.

Exercises 4–7 are multiple choice items. They refer to figure 14.19 which shows a transverse section of a leaf from a xerophyte. In each case you should choose ONE correct answer only.

4 Most stomata would be found at
   A 1
   B 3
   C 4
   D 6

5 Most moist air accumulates at
   A 2
   B 3
   C 5
   D 7

6 A structural feature possessed by some xerophytes but NOT shown in this diagram is
   A thick cuticle.
   B hairy surface.
   C rolled leaf.
   D stomatal pit.

7 This plant's natural habitat is most likely to be
   A an exposed moorland.
   B a salt-water mudflat.
   C a fresh-water pond.
   D a shady oak forest.

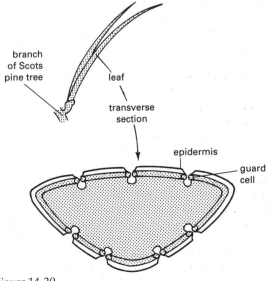

Figure 14.20

195

8 Figure 14.20 illustrates the leaf of Scots pine. This tree often grows on exposed windy hillsides.

a) Name THREE structural adaptations shown by the leaf that suit the plant to life in its natural environment.

b) Explain why the Scots pine tree does not need to shed its leaves annually in autumn.

9 Figure 14.21 shows a type of plant adapted to life in semi-desert environments which receive hardly any rain.

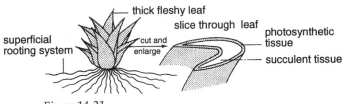

Figure 14.21

a) What general name is given to plants that can survive in desert conditions?

b) (i) Identify TWO structural features of this plant that enable it to make full use of rain on the rare occasions that it does fall.

(ii) Describe the role played by each of these features.

c) Predict TWO ways in which this plant's stomata would differ from those of a mesophyte.

10 Figure 14.22 shows a plant called sea plantain which is found growing in salt marshes round the British coast.

a) What general name is given to plants adapted to life in salt marshes?

b) (i) Predict what an analysis of the cell sap of this plant's root hairs would reveal about its water concentration compared to that of sea water.

(ii) In what way does this physiological factor adapt this plant to life in its natural environment?

c) In what way does the possession of fleshy leaves help this plant to survive?

11 Some halophytes such as glasswort are found living on tidal mud flats.

a) Would a shower of rain on an otherwise dry, breezy day make life easier or more difficult for such plants?

b) Explain your answer to a).

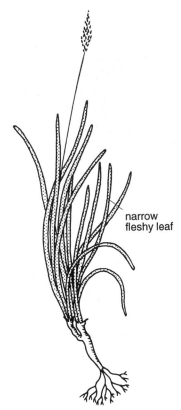

Figure 14.22

12 Figure 14.23 shows the stems of three different plants cut in transverse section. Although they are not drawn to scale, the diameter of each is given in the diagram.

plant type	letter in diagram	two structural features of stem that led to choice of letter
hydrophyte young mesophyte old mesophyte		

Table 14.3

b) Describe TWO ways in which the air spaces in stem B help the plant to survive in its natural environment.

Exercises 13 and 14 are multiple choice items. They refer to figure 14.24 which shows the pondweed *Potamogeton*. In each case you should choose ONE correct answer only.

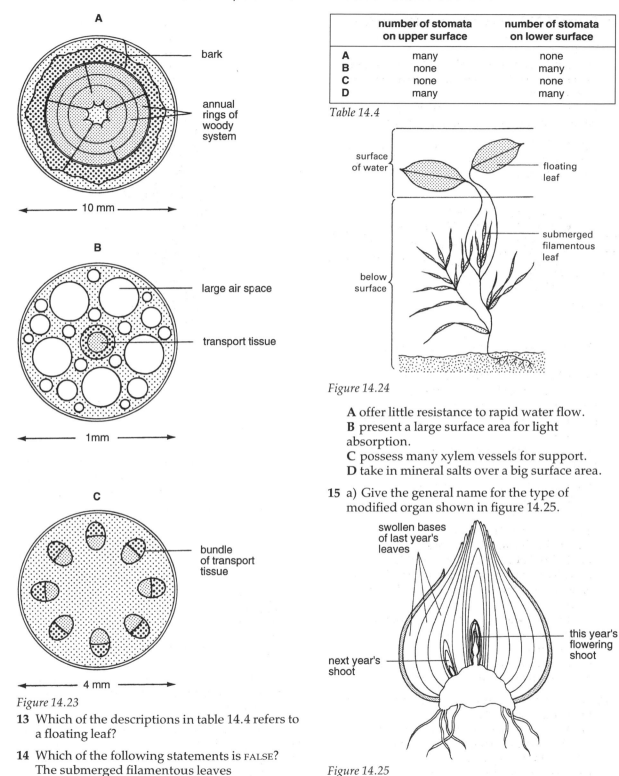

A

bark

annual rings of woody system

← 10 mm →

B

large air space

transport tissue

← 1mm →

C

bundle of transport tissue

← 4 mm →

Figure 14.23

	number of stomata on upper surface	number of stomata on lower surface
A	many	none
B	none	many
C	none	none
D	many	many

Table 14.4

surface of water

floating leaf

submerged filamentous leaf

below surface

Figure 14.24

A offer little resistance to rapid water flow.
B present a large surface area for light absorption.
C possess many xylem vessels for support.
D take in mineral salts over a big surface area.

13 Which of the descriptions in table 14.4 refers to a floating leaf?

14 Which of the following statements is FALSE? The submerged filamentous leaves

15 a) Give the general name for the type of modified organ shown in figure 14.25.

swollen bases of last year's leaves

this year's flowering shoot

next year's shoot

Figure 14.25

b) Name a flowering plant that makes this type of structure.

c) (i) Describe a feature of the metabolic rate of the cells in this structure that enables it to survive the winter without running out of food.

(ii) Describe how this feature increases the plant's chance of survival.

d) (i) Name the type of substance stored in the cells of this organ that enable the young plant to grow rapidly in the spring.

(ii) Why is the ability to grow rapidly in the spring of advantage to the young plant?

16 Figure 14.26 shows a soaked seed of an annual plant opened up after germination has begun.

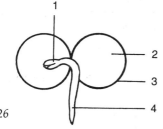

Figure 14.26

a) Name structures 1–4.

b) (i) Which of these structures contains stored food?

(ii) Which numbered structures make use of this food under natural growing conditions?

(iii) Why is it important that these structures use very little of the stored food during the winter months?

c) (i) At what time of the year is the food supply in the seeds of annual plants normally put to full use?

(ii) Why is the presence of a food store in the seed of great importance to the young plant?

(iii) How does the young plant survive once the supply of stored food in the seed is exhausted?

d) Apart from acting as a food store, the 'seed leaves' in the plant shown in the diagram play another role in helping it to survive. Name it.

17 Table 14.5 compares the percentage of solar energy reflected by four different plants.

a) Present the data as a bar graph with the bars arranged in ascending order.

b) Which plant is the (i) best (ii) poorest at reflecting light?

plant	% solar energy reflected
oak tree	4.3
prickly pear	9.4
marram grass	5.1
pond weed	2.9

Table 14.5

c) Broadly speaking, what relationship exists between the plant's efficiency at reflecting light and the availability of water in its natural environment?

d) Why do plants in desert environments need to be good reflectors of solar radiation?

e) Describe the structural feature of prickly pear that makes it an effective reflector of light rays.

18 Describe the means by which the compass plant manages to minimise its heat load at times of maximum insolation yet gain sufficient light for photosynthesis during much of the day.

19 *Encelia farinosa* and *Encelia californica* are related species of plant native to California, USA. One lives in moist environments and the other in the desert. One is covered by a layer of hairs which give it an overall silvery appearance. The other lacks hairs.

The graph in figure 14.27 shows the percentage of light of different wavelengths absorbed by the two species.

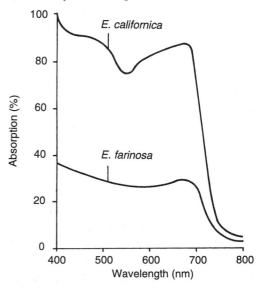

Figure 14.27

a) Which species (i) absorbed more light (ii) reflected more light?

b) (i) Identify which species has the silvery hairs and which is hairless.

(ii) Explain how you arrived at your answer.

c) (i) Which species is the native of the desert?

(ii) Explain your choice.

d) (i) Which of the two species will have the higher photosynthetic rate and show the greater annual yield of photosynthetic products?

(ii) Explain why.

Exercises 20–25 are multiple choice items and you should choose ONE correct answer only in each case.

Items 20, 21 and 22 refer to the graph in figure 14.28 which shows the amounts of two types of chemical present in young bracken leaves as they begin to grow in spring.

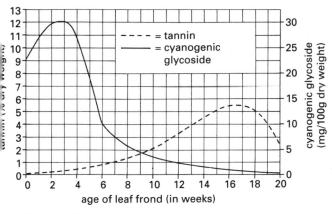

Figure 14.28

20 Which of the following conclusions can be correctly drawn from the graph?

A Younger leaves possess a high concentration of cyanogenic glycoside and a low concentration of tannin.

B Older leaves possess a high concentration of cyanogenic glycoside and a low concentration of tannin.

C Younger leaves possess a high concentration of both cyanogenic glycoside and tannin.

D Older leaves possess a high concentration of both cyanogenic glycoside and tannin.

21 The amounts of the two chemicals present in leaves at week 7 are

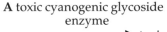

	tannin (% dry weight)	cyanogenic glycoside (mg/100g dry weight)
A	1.0	3.0
B	1.0	7.5
C	3.0	2.5
D	7.5	1.0

22 The concentration of cyanogenic glycoside shows its biggest drop in concentration between weeks

A 3 and 4   **C** 5 and 6

B 4 and 5   **D** 6 and 7

23 Which of the following equations correctly represents the process of cyanogenesis?

A toxic cyanogenic glycoside

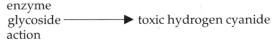

$\xrightarrow[\text{action}]{\text{enzyme}}$ toxic hydrogen cyanide

B non-toxic cyanogenic glycoside $\xrightarrow[\text{action}]{\text{enzyme}}$ toxic hydrogen cyanide

C toxic hydrogen cyanide $\xrightarrow[\text{action}]{\text{enzyme}}$ toxic cyanogenic glycoside

D non-toxic hydrogen cyanide $\xrightarrow[\text{action}]{\text{enzyme}}$ toxic cyanogenic glycoside

24 Which of the following is a modified side branch?

A hair   **C** sting

B spine   **D** thorn

25 Figure 14.29 illustrates a desert plant called *Echinocactus*.

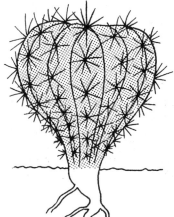

Figure 14.29

This cactus is protected from thirsty animals by possessing
A a rounded shape.
B a thick waxy cuticle.
C leaves reduced to spines.
D water stored in succulent tissues.

26 The graph shown in figure 14.30 gives the results of an investigation involving holly leaves. Twenty leaves were collected at each sample site and the average number of spines per leaf calculated.
a) What relationship exists between the average number of spines per leaf and the height of the leaf from the ground?
b) With reference to potential attacks by browing herbivores, explain why the leaves at 1 metre from the ground are well adapted to their particular environment.

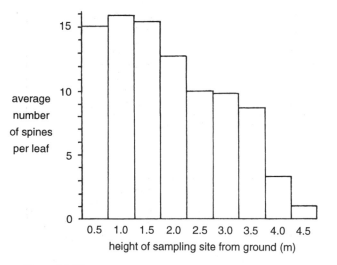

Figure 14.30

27 Figure 14.31 shows part of the rooting system of a mangrove tree. In an experiment the air in the region of submerged root immediately beneath each aerial root was sampled using a syringe.

This was done just before nightfall and then every aerial root on the plant was covered with a thick layer of grease. The air sampling procedure was continued over a period of 24 hours.

The untreated roots of another tree of the same species were used as the control. The results were graphed as shown in figure 14.32.

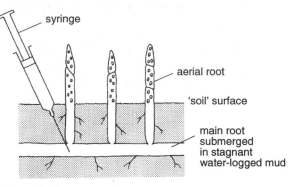

Figure 14.31

a) (i) In which submerged roots did the oxygen level show a steady drop and the carbon dioxide level a steady rise?
(ii) What physiological process in the root was responsible for these changes?
(iii) Why were these roots unable to obtain a renewed oxygen supply?
b) (i) In which submerged roots did the oxygen and carbon dioxide levels remain unchanged throughout the 24-hour sampling period?
(ii) Suggest why.
c) What does this experiment demonstrate about the route by which oxygen reaches the submerged roots of a mangrove tree?
d) Describe TWO features of an aerial root that make it well adapted to perform its function.

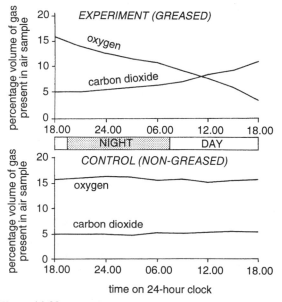

Figure 14.32

APPENDIX 1 THE GENETIC CODE

second letter of triplet					
	A	G	T	C	
A	AAA AAG AAT AAC	AGA AGG AGT AGC	ATA ATG ATT ATC	ACA ACG ACT ACC	A G T C
G	GAA GAG GAT GAC	GGA GGG GGT GGC	GTA GTG GTT GTC	GCA GCG GCT GCC	A G T C
T	TAA TAG TAT TAC	TGA TGG TGT TGC	TTA TTG TTT TTC	TCA TCG TCT TCC	A G T C
C	CAA CAG CAT CAC	CGA CGG CGT CGC	CTA CTG CTT CTC	CCA CCG CCT CCC	A G T C

(first letter of triplet — left; third letter of triplet — right)

(A = adenine, G = guanine, T = thymine, C = cytosine)

Table Ap 1.1   DNA's bases grouped into 64 (4×4×4) triplets

abbreviation	amino acid
ala	alanine
arg	arginine
asp	aspartic acid
aspn	asparagine
cys	cysteine
glu	glutamic acid
glun	glutamine
gly	glycine
his	histidine
ileu	isoleucine
leu	leucine
lys	lysine
met	methionine
phe	phenylalanine
pro	proline
ser	serine
thr	threonine
tryp	tryptophan
tyr	tyrosine
val	valine
●	chain terminator

Table Ap 1.3   Key to amino acids

codon	anti-codon	amino acid	codon	anti-codon	amino acid	codon	anti-codon	amino acid	codon	anti-codon	amino acid
UUU	AAA	} phe	UCU	AGA	} ser	UAU	AUA	} tyr	UGU	ACA	} cys
UUC	AAG		UCC	AGG		UAC	AUG		UGC	ACG	
UUA	AAU	} leu	UCA	AGU		UAA	AUU	} ●	UGA	ACU	●
UUG	AAC		UCG	AGC		UAG	AUC		UGG	ACC	tryp
CUU	GAA	} leu	CCU	GGA	} pro	CAU	GUA	} his	CGU	GCA	} arg
CUC	GAG		CCC	GGG		CAC	GUG		CGC	GCG	
CUA	GAU		CCA	GGU		CAA	GUU	} glun	CGA	GCU	
CUG	GAC		CCG	GGC		CAG	GUC		CGG	GCC	
AUU	UAA	} ileu	ACU	UGA	} thr	AAU	UUA	} aspn	AGU	UCA	} ser
AUC	UAG		ACC	UGG		AAC	UUG		AGC	UCG	
AUA	UAU		ACA	UGU		AAA	UUU	} lys	AGA	UCU	} arg
AUG	UAC	met	ACG	UGC		AAG	UUC		AGG	UCC	
GUU	CAA	} val	GCU	CGA	} ala	GAU	CUA	} asp	GGU	CCA	} gly
GUC	CAG		GCC	CGG		GAC	CUG		GGC	CCG	
GUA	CAU		GCA	CGU		GAA	CUU	} glu	GGA	CCU	
GUG	CAC		GCG	CGC		GAG	CUC		GGG	CCC	

(U = uracil)

Table Ap 1.2   mRNA's 64 codons, tRNA's anticodons and amino acids coded

α-amylase 146
abscissic acid (ABA) 149, 150, 163
abscission 144, 149, 150
   layer 149
acetyl coenzyme A 38, 39
actin 3
activation energy 16
active site 17
adenine 49
adenosine diphosphate (ADP) 37
adenosine triphosphate (ATP) 37–43
adventitious root 144
aeration tissue 187, 194
aerial root 194
aerobic respiration 37, 40
albinism 71
albino 71
aleurone layer 147
allele 80, 105
   dominant 105
   multiple 115
   recessive 105, 108
amino acids 1, 58, 97
   key to 201
amylase 16, 17
anaerobic respiration 37, 40
anaphase 78
anticodon 60
aphid 131
apical dominance 144
autoadiograph 132
auxin 142–3
   effect with gibberellin 146
   effect with cytokinin 148, 149

bark 131
barley grain 146
base 49, 58
   pair 49
   triplet 58
   β-galactosidase 66, 67, 68
blood group 116
bond
   chemical 49
   covalent 10
   disulphide 3
   hydrogen 1, 49, 51, 53
   peptide 1
bulb 189, 190

callus 149
Calvin cycle 42

carbohydrate 5, 122, 123
carbon 1, 5, 9
   fixation 41, 42
carbon dioxide 39, 43, 122, 126
   acceptor 43
carrier 114
catalyst 16
cell
   guard 128
   membrane 3, 4, 11
cellulose 9, 43, 142
centromere 65, 76
characteristic 105
chemical bond 49, 51
chemical pathway 37
chiasma 76, 78, 81
chlorophyll 41
chloroplast 41
cholinesterase 24
chromatid 78, 88
chromosome 49, 88, 95, 110
   complement 75
   deletion 88
   duplication 88
   homologous 75, 88, 90, 95, 110
   inversion 88
   number 91
   sex 93, 112
   translocation 89
citric acid 38, 39
clinostat 162
codon 58, 59, 61
coenzyme 37
co-factor 18, 19
coleoptile 142, 161
collagen 3
colour blindness 114
companion cell 139
compass plant 192
condensation reaction 1, 6, 8, 10
conjugated protein 2, 4
conservation
   of base sequence 51
corm 189, 190
cotyledon 190
covalent bond 10
cri du chat syndrome 89
crista 37
cross 105–8
   back (test) 108, 110
   dihybrid 105
   monohybrid 105

cross pollination 170
crossing over 78, 81, 88, 110
cross-over frequency 80, 81
cuticle 191
cyanide 192
cyanogenesis 192
cystic fibrosis 99
cytochrome 39
cytokinesis 148
cytokinin 148, 149
cytosine 49

day neutral plant 168
deletion
   chromosome mutation 88
   gene mutation 70
denaturation 19
deoxyribose 49
differentiation 148
dihybrid cross 105
diploid 75
disaccharide 6
disease resistance 96
disulphide bond 3
DNA (deoxyribosenucleic acid) 49, 50, 58, 65, 67
   genetic code 201
   polymerase 51
dominance
   apical 145
   incomplete 116
dormancy
   bud 146, 149, 170
   seed 149, 150, 167
double helix 49
Down's syndrome 91
drought 185
duplication
   chromosome mutation 88
   gene mutation 69
dysfunction 98
   biochemical 99

early prophase 78
Echinocactus 186
elastin 3
Elodea 123, 124, 125
embryo 147
end product 17
endosperm 146
energy 122, 126, 133
   activation 16

enzyme 16–24, 51, 58, 69
  action 16
  control of biochemical reaction 68, 99
  fruit ripening 150
  structure 3
Escherichia coli 66
ethanol 40
ethylene 150, 151
eucalyptus 192

factor VIII 99
fat 10
fatty acid 10
fertile organism 95, 96
fibrous protein 2, 3
florigen 169
fluis-mosaic model 3, 4
food storage 6

galactose 66
gamete 89, 105
  diploid 90, 95
  haploid 90, 95
  mother cell 78
gene 65, 88
  action 66
  expression 65
  linked 110
  mutation 69
  operator 66–7
  promoter 66–7
  regulator 67
  sex-linked 113
  structural 66–7
  therapy 99
gene operon theory 66–7
generation
  first filial (F_1) 105, 107
  second filial (F_2) 105, 107
genetic
  code 58, 97, 201
  instruction 58
genome 94, 95, 96
genotype 105
geotropism 163
Gibberella 145
gibberellic acid 145, 167
gibberellin 145
  effect on dormancy 146, 147
  mechanism of action 145, 146
glasswort 186
globular protein 2,3
glucose 5, 40, 66, 126
  α and β 5

glycerate phosphate (GP) 43
glycerol 10
glycogen 6, 7
glycolysis 37
glycosidic bond 6
grana 41
growth substance 142
guanine 49
guard cell 128

haemoglobin 4
  normal 97
  S 70, 98
haemophilia 99, 115
hair 191
Hakea 185
halophyte 186, 187
haploid 75
herbicide 144
homologous pair 80
hormone
  plant 142
  summary 151
humidity 130
hybrid
  sterile 95, 96
  vigour 96
hydrogen 1, 5, 9
  bond 1, 49, 51, 53
hydrophilic 11
hydrophobic 11
hydrophyte 187, 188

incomplete dominance 116
independent assortment 79
indole acetic acid (IAA) 142
inducer 67
inhibitor 22
  competitive 23
  irreversible 24
  non-competitive 24
insertion
  gene mutation 70
insolation 191
insulation 191
internode 145
interphase 78
inversion
  chromosome mutation 88
  gene mutation 70

karyotype 92, 93, 94
keratin 3
Klinefelter's syndrome 94
Krebs' cycle 38–40

lac operon 66–7
lactose 66
lag phase 66
lamella 41
leaf fall 149
lenticel 194
light 130
  energy 41, 122
  wavelength 125
limiting factor 123, 124
linkage 80, 111
linked genes 80, 110
lipid 10
lock-and-key 17, 18
locus 65, 115
long day plant 168, 169
low oxygen habitat 193, 194

maltose 6
mangrove 194
marram grass 185
matrix 37
meiosis 75–8, 88
  non-disjunction 89, 93, 95
melanin 71
Mendel 105
mesophyte 184
messenger RNA (mRNA) 53, 59, 60, 61
  transcription 142
metabolic
  activity 133
  pathway 69
metabolism 68
metaphase 78
mitochondria 37
mitosis
  non-disjunction 90
monohybrid cross 105
monosaccharide 5
mucus 99
multiple alleles 115
mutation 97, 99
  gene 69
myosin 3

NAD (H_2) 37, 39
NADP (H_2) 42
nastic response 164
nitrogen 1
non-disjunction 89, 90, 95
  of sex chromosomes 93
nuclear membrane 78
nucleic acid 49
nucleotide 49–53, 65
nucleus 53

oak tree 184
operator gene 66–7
operon 66
orientation of leaf 192
oxaloacetic acid 39
oxygen 1, 5, 9, 38, 122, 127
    debt 40

parthenocarpy 144
peptide bond 1, 2, 58
perennating structure 170
pH 20
phenotype 105
phenylalanine 71
phenylketonuria (PKU) 71
phloem 130, 131, 169
phosphate 10, 49
phospholipid 10, 11
phosphorylase 26
phosphorylation 37, 39
photolysis 41, 42
photonasty 164
photoperiod 168, 170, 171
photoperodism 168
    mechanism 168, 169
photophosphorylation 42
photosynthesis 41, 43, 122, 128
    rate of 123
phototropism 161, 162
phytochrome 166, 167
    mediated response 168–71
polymerisation 51, 53, 61
polypeptide 58, 61
    chain 1, 2
polyploidy 94, 95
    economic significance 96
polysaccharide 6
pore 194
potometer 129
predation
    adaptations against 192
prickly pear 191
promoter gene 66–7
protein 1, 97
    structure 58
    synthesis 58, 59
Punnett square 105, 108
pyruvic acid 37, 39, 40

R group 1
radioactive tracer 132

random assortment 80
recombinant 108, 110
recombination 108
reflective layer 191
regulator gene 66
replication 51
repressor 66
respiration 37, 126, 128
    rate of 127
respirometer 127
ribonucleic acid (RNA) 50
ribosome 60, 61
ribulose bisphosphate (RuBP) 42
ringing 131
ripening 151
RNA 50
    messenger 53, 59
    polymerase 53, 67
    transfer 59
    triplet 201
root
    adventitious 144
    aerial 194
    tuber 188
rooting powder 144

seedless fruit 144
senescence 149
sex
    determination 112, 113
    linkage 113, 114
    ratio 112, 113
sex-linked gene 113
short day plant 168
sickle cell anaemia 70, 98
sieve plate 131
sieve tube 131
sleep movement 164
spectrum 125
spindle fibre 78
spine 185, 193
starch 6, 7, 8, 43
sterile organism 95
sting 193
stoma 128, 184
    sunken 185
stroma 41
structural gene 66–7
structural protein 3
substitution
    gene mutation 70

substrate 17
succulent tissue 186, 187
sundew 165
syndrome
    cri du chat 89
    Down's 91
    Klinefelter's 94
    Turner's 93

telophase 78
temperature 124, 127, 130, 164
tetraploid 95, 96
thermonasty 164
thorn 193
thymine 49
touch 165
tracer 132
transcription 53, 67
transfer RNA (tRNA) 59, 60
translation 60, 61
translocation
    chromosome mutation 89
    in phloem 130–3, 169
transpiration 128–130
    pull 129
triose phosphate 43
triploid 95
tropic response 161, 162
turgor 129
Turner's syndrome 93
turnover number 22
tyrosine 71

variable factor 125
variation 81

water 122, 126
    soil 130
water-lily 188
water-milfoil 187
wax 191
weedkiller, selective 144
winter bud 170

xerophyte 185, 186
xylem 185

zygote
    diploid 75
    tetraploid 95
    triploid 95